MW01259029

MICROORGANISMS IN THE BIBLE

MORLEY, MISSOURI

MICROORGANISMS IN THE BIBLE

MICROORGANISMS IN THE BIBLE

To Chris
Best of luck
Larry Elliott

Chris - thought of you and our time @ WKU when I saw this book.
Barb

Larry P. Elliott, Ph.D.

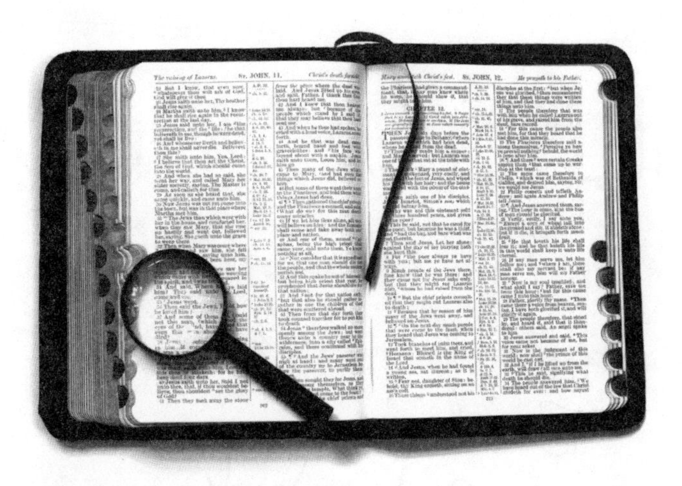

AP™
Acclaim Press
— Your Next Great Book —
P.O. Box 238
Morley, MO 63767
(573) 472-9800
www.acclaimpress.com

Book Design: Devon Burroughs
Cover Design: M. Frene Melton

Library of Congress Cataloging-in-Publication Data
Elliott, Larry P., Ph. D.
Microorganisms in the Bible / by Larry P. Elliott, Ph.D.
pages cm
ISBN-13: 978-1-938905-25-4 (alk. paper)
ISBN-10: 1-938905-25-3 (alk. paper)
1. Microorganisms. 2. Nature in the Bible. I. Title.
BS660.E45 2013
220.8'579--dc23
2013007739

First Printing: 2014
Printed in the United States of America
10 9 8 7 6 5 4 3 2 1

Contents

Acknowledgments 6
Preface 7
Microbes in the Beginning 11
The Flood 14
Wine 16
Beer 20
Bread 23
Vinegar 26
Olives 29
Cheese & Butter 32
Pickles 35
Flax 37
Coral 39
Spoilage of Grains & Grain Products 42
Diseases 45
Biblical Methods Used to Control Microbial Growth 74
Embalming 88
Physicians & Medicine 89
Sanitary Code 91
Is the Dead Sea Dead 93
Manna 95
Normal (Indigenous) Microflora of Humans 97
Oil 103
Conclusion 107
Glossary 109
About the Author 117
Index 120
Scriptural References 125

Acknowledgments

I first become interested that the Bible contained information concerning microbes when I took a microbiology course taught by Olive Thomas at William Jewell College. While earning my Master's degree and Ph.D. in bacteriology at the University of Wisconsin, Madison, I was too busy to write a book concerning this topic. I thought surely the inspiration to write this book would come some time during the 37 years I taught microbiology and related courses at Western Kentucky University in Bowling Green, Kentucky, but it did not. Not until my second year of optional retirement in the biology department did the time finally come for me to write this book. I have collected this material over the years and have finally got it organized.

My family has been a blessing to me. In fact, this book is written in memory of my mother Margaret Marie Elliott and father Melvin J. Elliott. He was a Christian, a deacon in our church, and a very good role model. He was an inspiration to everyone that knew him. My mother and father were great parents; if I could have picked out my parents I would choose them both again. I was fortunate to be raised in a Christian home. Particularly because of their influence, I accepted Jesus Christ as my Savior at the age of nine. My wife Wilma, whom I met at William Jewell College, has been an excellent and supportive wife since 1961 when we married. My daughters, Kerrie Lynn, Kimberly Ann, and Kelly Jo, all graduates of Western Kentucky University, continue to add meaning to my life. Kerrie McDaniel has a Ph.D. in plant molecular biology and has blessed our family with two children, Christopher and Sarah. Kimberly has a Master's Degree in nursing and is married to Rett Dallas. They have blessed us with their two children, Carder and Aby Jewell. Kelly is an M.D. specializing in Pediatrics and is married to Shawn Kries. They have two children, Audry and Eli. Since they are the youngest of the grandchildren they are being spoiled by us all, grandparents, aunts, uncles and cousins. We all continue to strengthen our love for each other, which bonds us more as each day goes by.

I am also deeply indebted that Dr. Rollin Burhans was our pastor at First Baptist Church in Bowling Green for ten years. He had a wonderful helpmate in Delma. They both have passed away and are missed very much. If Baptists chose saints, then Dr. Rollin and Delma would be chosen.

Last, I have been blessed with a multitude of students taught in both Sunday School and at the university, who have challenged me and helped make my life challenging and fun. One of my Master's degree students, Mark Clauson, who continues to teach in the Western Kentucky University department of Biology, deserves special mention for his encouragement and assistance in bringing this book to fruition. Also friends and students from grade school years until now have aided in shaping my life to be an honest and contributing person to society.

Preface

The Holy Bible is a book of scripture including the Old Testament and New Testament, which is the Word of God. It is a record and interpretation of God's self-disclosure to man. It is an authentic account of the revelation of God in Jesus Christ for the redemption of man.

To be understood, the Bible must be seen as literature of different kinds and forms. For example, one can find examples of history, law, poetry, drama, prophecy, wisdom literature, apocalyptic literature, hymns, anthems, sermons, statistical lists, parables, and stories within its pages.

Several books have been written that discuss the Bible and how it relates to nature, on such topics as fauna and flora, animals, plants, flowers, trees and birds. It seemed apparent to me that a book concerning microbes in the Bible would be beneficial. There is literature concerning diseases mentioned in the Bible, but no concerted effort has been made pulling this literature together until now.

It is not my purpose to mention every verse in the Bible that mentions a microbe (yeasts, for example), but to cite a few verses that document they were present during biblical time. Obviously at times the microbe mentioned in the bible will be speculative, but I will present which microbes would be involved in the actions mentioned (such as the microbe responsible for food spoilage, etc.).

Also, I will mention possible antimicrobials that were used in the Bible and the type of microbial reactions they inhibit. This book assumes little or no scientific background. Hopefully we will understand that this ancient book, The Bible, has well documented the existence of microbes upon the earth and that this information can be used by many people across many different disciplines. Microbiology has always been fun for me to study. Hopefully, this book will enable you to enjoy microbiology as the microbial world is mentioned in the bible.

I would enjoy hearing from you. How did this book work for you? I would like to hear from you if some material is lacking or inaccurate in any way.

Contact me at:

Western Kentucky University
1 Big Red Way
Biology Department
Bowling Green, KY 42101-3576

Or by e-mail:

Larry.Elliott@wku.edu

Warmest regards,

Larry P. Elliott

MICROBES IN THE BEGINNING

Genesis chapter 1 gives us a glorious picture of God's creative acts and His charges and promises to humankind. Creation is viewed as having occurred over an indefinite period of time, and having proceeded from the lower forms to the higher.

It does not appear that the days in Genesis 1 can be held to 24 hours. In 2:4b the entire period of creation is referred to as a day. The sun, which is responsible for the 24-hour day, was not in its place until the fourth day.

Was there a Big Bang in the creation of the universe? DeYoung strongly suggests in his book "Astronomy and the Bible" that there are chronological discrepancies between the Bible and the Big Bang Theory. The Big Bang can be defined as a random chance event where instability supposedly developed in the original "kernel" of mass

Replica of the Leeuwenhoek microscope

energy. Scriptures rule out such an accidental explosion that led to a rapid expansion of space that carried matter along with it. Regardless of contemporary scientific views on the beginning of our universe, the book of Genesis tells us that Gods hands were at work.

Surely, the first organisms created were bacteria. To the Biblical writer, microbes were too remote and unknown to require any detailed attention. In fact it was Anton van Leeuwenhoek in 1675, who was the first person to observe microbes. He accomplished this with a simple microscope he made himself, leaving little doubt that he observed bacteria, fungi and many forms of protozoa.

Bacteria are among the most successful organisms on Earth. Organisms like these are termed prokaryotes, a name that reflects the absence of a nucleus (karyo is Greek for "nut" [or nucleus] and pro means "primitive"). Eukaryotic cells are those that make up humans, eu (true) and karyhon (nucleus). These two types of cells are different in other ways and this detail can be examined in current microbiology texts.

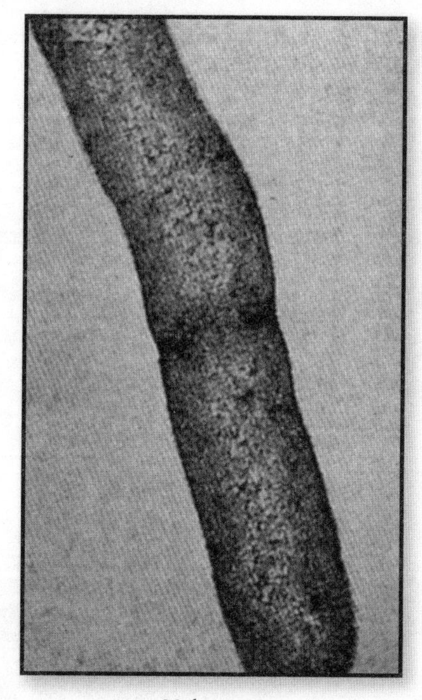
Methanogens

Prokaryotes differ enough to be split into two large groups called domains. A domain is the highest taxonomic category, higher than kingdom. Bacteria are in the domain Archaea and Bacteria. All other organisms are in the domain Eukarya, including humans.

Most bacteria on this planet are members of the domain Bacteria. The archaebacteria are particularly found in harsh environments. They can live under extremely acidic and extremely hot conditions. One archaeon is found in hydrothermal vents at the ocean's bottom. The methanogens produce methane under oxygen-free conditions (anaerobic) in volcanic rock, marshes, lake bottoms, and animal feces. Others are halophiles that live in high-salt environments, such as Utah's Great Salt Lakes.

In the beginning God established laws of physics and chemistry and brought the earth to a certain place and then put it into man's charge. Whatever the first microbe was, deep down at the level of the essential biochemistry of life,we find the same proteins control cell chemistry, and the same nucleic acids carry and reproduce the hereditary material.

Additional Reading:

Dye, D.L., *Faith and the physical world: a comprehensive view.* William B. Eerdmans Publishing Co., 1966

Lederberg, J., *Encylopedia of Microbiology*. San Diego: Academic Press, 1992.

The Flood

In Genesis 6:1-9:2 it states that as a result of the wickedness of the world, God decided to destroy his creation—man and beast, reptiles and birds. Just as in the Garden of Eden the serpent tempted Eve, new deviant divine beings corrupted mankind.

Noah was righteous and blameless in a generation where such characteristics were almost unknown. God did not take him because God had work for him to do. God told him to make an ark because God was going to destroy His creation—man and beast, reptiles and birds.

Since the flood was of the earliest ages, it covered the whole earth and left its mark on every continent on the planet. Into the ark Noah was to take all his house and seven pairs, male and female of all clean beasts and birds, for later they would be used for sacrifice and food. He also took one pair, male and female of all unclean beasts. Since the animals that lived in the water did not have to be brought aboard, about 300 species of beast and birds were collected and placed in the ark. The task was lightened since God brought the animals to Noah; he did not have to round them up (v. 20b).

In 7:16, states the "Lord closed the door on him" including his sons Shem, Ham and Japheth, Noah's wife and the three wives of his sons demonstrated that God was casting his divine care on them. How else could all the animals live together, shut in the ark floating about for five months without some of them dying of disease? This is especially so considering modern sewage disposal and water purification was not available. The intestinal microbes of both man and animals could have been spread from feces to fingers to mouth. It was miraculous that any life emerged from the ark. This is so since zoonotic diseases are diseases by infectious agents that can be transmitted between (or are shared by) animals and humans.

There are many microbial diseases that particularly the animals could have spread to Noah and his family. Presently, over 150 zoonoses are known. In terms of historical significance, anthrax, bubonic plague, and brucellosis rank among the classical diseases that occur in humans and animals. Bats can transmit diseases such as rabies and histoplasmosis.

One must note, however, that at the end of the time period, Noah and his family and all the animals came out on dry land. Obviously, supernatural intervention was involved.

Additional Reading:

Batzing, B.L., *Microbiology, an Introduction,* Brooks/Cole, Thomson Learning, 2002.

WINE

In Genesis 9:20 the first mention of wine is made where it states that Noah was a husbandman, and planted a vineyard and drank of the wine he made and became drunk. The Bible speaks of wine a hundred and forty-one times. It is not my purpose to document all these verses but to relate many of the verses that describe wine or that are closely related to the subject.

Wine most often refers to the beverage made from grapes, but can be made from other fruit juices. It is a moderately alcoholic beverage (12% or so) made from the juices of fresh, ripe grapes. The soil of Palestine is favorable for the cultivation of grapes (Num. 13:20-24). The vines were generally cultivated on hills (Ps. 80:10, Amos 9:13), protected by walls (Num. 22:24), and were provided with a tower for watchmen (Mark 12:1). After gathering the grapes, they were spread in the sun before being pressed into juice or must by barefoot vintagers (Isa. 63:2, Rev. 14:19-20; 19:15). Generally several men trod on the grapes together (Neh. 13:15; Jer. 25:30, 48:33). The winepress where the grapes were trodden usually were of stone, hallowed out of a flat rock, from which juice flowed through a hole near the bottom into a vat (Neh. 13:15; Isa. 63:3; Matt. 21:33; Isa. 1:8).

The juice, being acidic and rich in sugars (particularly glucose and fructose), is an ideal medium for the growth of yeasts. For red wine, black grapes are used, including skins and stalks. It certainly must be red wine the Hebrews made because the Bible speaks of the redness of wine and does not mention white wine. The waxy bloom on the surface of grapes contains many different microbes, including the wine yeast *Sacharomyces cerevisiae var. ellipsoideus.* Yeasts are a type of fungus that is unicellular and not characterized by mycelia, which are a mass of threadlike filaments. Yeast cells may be round, ovate or elongated. They multiply asexually by multilateral budding and sexually by producing ascospores in asci. These top yeasts are very active fermenters and grow rapidly at 20° C. The clumping of cells and rapid production of carbon dioxide sweep the cells to the surface, hence the term top yeasts.

After about six hours of fermentation, the yeast converts the grape juice into ethyl alcohol and carbon dioxide gas. After completion when the yeasts have used up most of the sugar or are inhibited by the amount of alcohol they have generated, which takes from a few days to a couple of weeks, it is transferred to jars (Jer. 13:12; Josh. 9:4, 13) or skins for further fermentation and storage. The skins were usually whole goat hides, sewed up with hair outside with the neck and feet tied. Also the skins were made of tanned goat's hide, and they had wooden stoppers. The freshly made wine was transferred to new wine skins since the old ones would burst under the pressure of the gases produced (Matt. 9:17; Mark 2:22; Luke 5:37-38). The Hebrews filtered their wine before drinking to remove yeast cells and other sediment, probably even insects (Matt. 23:24). They drank their wine out of metal goblets or earthenware mugs (Song of Sol. 7:2; Esther 1:7). The Greeks and Romans scented their wine with thyme, cinnamon, roses and jasmine flowers, and this habit spread to the richer classes among the Jews (Prov. 9:5; Song of Sol. 8:2).

Additional uses of wine were for gifts (1 Sam. 25:18; 2 Sam. 16:1; Deut. 15:14), as a tonic to revive persons who fainted (2 Sam. 16:2), and for medicinal purposes. It was titheable (Deut. 18:4). It was used in place of water, which generally was impure. Paul counseled Timothy to use wine in place of water for his stomach's sake (1 Tim. 5:23). Wine was also used to dress wounds (Luke. 10:34).

Temperance and moderation appeared to be the prevalent idea toward wine in Bible times. New wine was to be used by maidens (Zech. 9:17). It was used for pleasure and cheering qualities (Ps. 104:15; Prov. 31:6-7; Ecc. 9:7; 10:19). It was used in the sacrificial meal (Deut. 14:23:26; 1 Sam. 1:24; Amos 2:8). Jesus made water into wine at Cana (John 2:7-11). The fruit of the vine was used in the Last Supper (Matt. 26:29).

In Old Testament times wine was not diluted, but in New Testament time it was mixed with water and thus had low alcoholic content. The Greek word translated wine to mean grape juice as well as fermented juice. Many New Testament scholars, however, believe the wine was fermented.

The Bible is full of warnings in the abuse of wine (Prov. 20:1; Isa. 5:11-12; Hosea 4:11; Amos 6:6). In Prov. 21:17 it states that a man who loves wine and oil will never grow rich. Also one should not keep company with drunkards or those who are gluttonous (Prov. 23:20). Prov. 23:29-35 relates the problems of intoxication: red-eyed staggering, dizzy, vomiting, mind confused, steps wandering, quarrelsomeness, and himself as helpless as a mariner asleep in the midst of the ocean.

Other dangers of wine could include alcoholism. Remember Noah and the problem drinking gave him (Gen. 9:21-24). Lot, while drunk, lay with his daughters and each bore a son by their father (Gen. 19:31-38). Others mentioned as excess drinkers are: Elah's Naabal (1 Sam. 25:36-37), Benhadad the king of Syria, and thirty-two kings who were probably tributary princes (1 Kings 20:12-16), King Ahasuerus (probably Xerxes of Persia) who drank for seven days (Esther 1:7-10), and the Ephramites (Isa. 28:1). Judah and particularly Israel, the subject of Amos' prophecies, drank wine bought with money of those whom they unjustly fined (Amos 2:8).

Over-indulgence drinking wine was warned against (1 Tim. 23:8; Titus 2:3). Total abstinence from wine was practiced by the Nazarites (Num. 6:3; Jer. 35:6). Famous Nazarites included Samson (Judges 12:7, 24) and his father Manoah (Judges 13:4, 7, 14). Jonadab, son of Rechab, was charged not to drink wine and his descendants obeyed (Jer. 35:5-6, 14) and they were zealous for God (2 Kings 10:15-23). The Rechabites were a nomadic tribe belonging to the Kenites of Hemath (1 Chron. 2:55) of the family of Jethro, or Habab, Moses' father-in-law (Num. 10:29, Judges 1:16). Since drinking often led to sin, priests were forbidden to drink wine while on duty (Lev. 10:9, Ezek. 44:21). It is interesting that Daniel determined not to contaminate himself by touching the food and wine assigned to him by the king of Babylon (Daniel 1:8), but ate only vegetables and drank water (Daniel 1:12-16, 10:3). John never touched wine or strong drink (Luke 1:15).

Many Christians like myself basically do not believe in social drinking. It is difficult, however, to find proof texts to support the teetotalers view. One of the best texts is in Romans 14:21, where it states that is not good to eat flesh, nor to drink wine, nor do any thing that makes our brother blunder or makes him offended. For me the Nazarite views are good enough to follow; they could see the evil of alcohol even then. If we are to learn from history, all we have to do is to study the harm the liquor industry has done to the total world today. Since we have the benefit of history it should be easier for us to abstain from drink because, like the Nazarites, we know alcoholic beverages inflame the passions, intoxicate the brain, and create a taste for luxurious indulgence.

Our bodies are meant to be God's temple and consequently, we have a moral obligation to keep it as healthy as possible.

Additional Reading:

Prescott, L.M., J.P. Harley, and D.A. Klein. *Microbiology, 4th Edition*, WCB McGrae-Hill, 1999.

BEER

Unlike wine, beer is not mentioned in the King James Version of the Bible, but in other translations such as New International Version and International Standard Version, strong drink is translated as beer. In Prov. 31:6 it states, "Give strong drink unto him that is ready to perish, and wine unto those that be of heavy hearts." Prov. 20:1 states, "Wine is a mocker, strong drink is raging; and whosoever is deceived thereby is not wise." In Prov. 28:7 we read, "But they also have erred through wine and through strong drink are out of the way; the priest and the prophet have erred through strong drink, they are swallowed up of wine, they are out of the way through strong drink; they err in vision, they stumble in judgment". In Judges 13:4 it talks about a woman who was barren and an angel told her not to drink wine nor strong drink, and not to eat unclean things. In v. 7 the angel said she would conceive and bear a son and continue to drink no wine nor strong drink. In v. 24 we notice that she had a son and he was called Samson. Similar verses, appear in Lev. 10:19, Num. 6:3, Deut. 29:6, Is. 5:11, Is. 29:9, Mic. 2:11 and Luke 1:15. In all these passages strong drink obviously is not wine. It is interesting that the Bible view on strong drink is negative.

The first reference to beer dates to 6,000 B.C. where it is written on clay tablets that the Babylonians had a recipe for beer brewed from barley. Records show that the Egyptians were brewing beer in 4,000 BC. In 1996 a report concerning Egyptian beer was made where a scientist examined beer residue and found they used barley to make malt and did not use hops in brewing.

Since the Hebrews lived in Egypt for 400 years it is apparent that they learned the art of brewing. The word beer is derived from the Anglo-Saxon baere, meaning barley. The exact procedures that were used to make beer in ancient times are unknown. Thus, a modern production scheme will be given. To make beer, cereal grains (usually barley) are malted (partially germinated) to increase the concentration of digesting enzymes (amylase) that provides the sugar for fermentation. This is done in a malting tank. The malt is ground with hot water (about 60° C) to produce mash. The soluble materials are drawn off and can be used for cattle feed. This is done in a mashing tank. Brewers then remove the liquid portion called wort. The wort is now boiled to inactivate the enzymes. This wort now flows into the hopping tank. Dried petals from the Hops plant (*Humulus lupulus*) are added to the wort for flavoring and the mixture is boiled to stop enzymatic action. Hops also give it color and stability. The bitter flavor in beer comes from the hops extract. Hop leaves also contain two antimicrobials acting as a preservative in the beer. The fluid is then filtered and transported to the fermentation tank. Adding yeast in large quantities is called pitching; about 10-30 pounds of yeast is added to a one hundred barrel vat. The type of yeast used depends upon the beer desired. Stout or ale requires "top", which are yeasts that ferment between 50° and 75° F and generally goes for three to five days. The yeast used is *Sacchromyces cerevisiae,* which gives a uniform dark cloudiness in beer and is carried to the top of the vat by forming carbon dioxide. German and most American lagers use "bottom" yeast that grows best between 40° and 50° F and the fermentation goes between eight to sixteen days. The "bottom" yeast used is *Sacchromyces carlsbergensis* and the fermentation is slower and produces a lighter, clearer beer with less alcohol content. This yeast doesn't clump and settles to the bottom. About three-quarters of the world's beer is lager beer. During fermentation most of the sugars are converted to alcohol and carbon dioxide. After three to five days for the top yeast fermentation and eight to sixteen days for the "bottom" yeast fermentation the beer is run into storage vats for aging and takes six weeks to six months depending on the type of beer being made. After this, the beer is filtered and run into storage tanks from which packaging occurs.

Of course, the Hebrews did not have this technology but indeed produced beer. Their technique probably was quite simple. Certainly barley grains were available to them for yeast fermentation. In Ezek. 4:9 Ezekiel was instructed by God to put wheat, barley, beans, lentils, millet and fitches into one vessel, this is what the people would eat during the siege of Jerusalem. In Num. 5:15 barley is used for an offering. In Deut. 8:8 it states that the promised land shall contain, wheat, barley, vines, fig trees, pomegranates, a land of olive oil and honey. Probably the Hebrew beer was made like the Egyptians where they crumbled lightly baked leavened bread into water. This contains the yeast *Sacchromyces cerevisiae*. This mixture was put into a large vessel, now called a vat. This bread water mixture that probably contained *S. cerevisiae* was allowed to ferment; it could then be flavored with honey or date juice.

Beer's mention in the Bible is scant to that of wine (over 200 occurrences), but it seems clear to me that wine and strong drink are two separate drinks.

Additional Reading:

http://www.discerningtheworld.com/2012/01/04/drinking-beer-and-wine-the-bibles-counsel

http://www.neatorama.com/2009/02/18/neatolicious-fun-facts-beer

BREAD

The use of yeast as a leavening agent dates back to the Stone Age, though the purpose of this fermentation is not the production of alcohol but carbon dioxide to raise or leaven bread. The strains of *Saccharomyces cerevisiae* used today came originally from the brewery industry. In baking, carbon dioxide forms the bubbles of leavened bread. Aerobic conditions favor carbon dioxide production and are encouraged as much as possible. This is why the bread dough is kneaded repeatedly. Whatever ethanol is produced evaporates during baking.

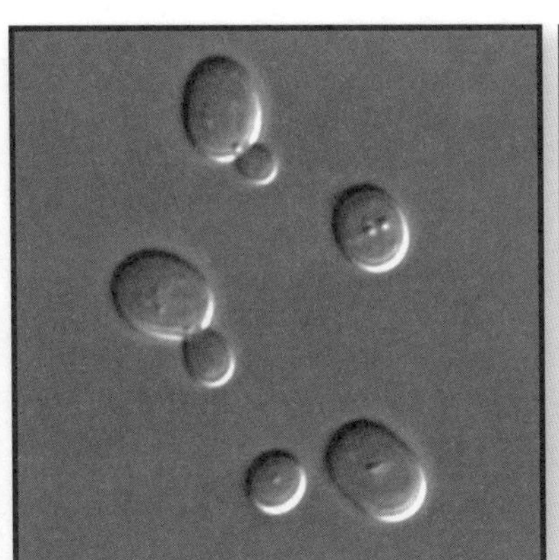

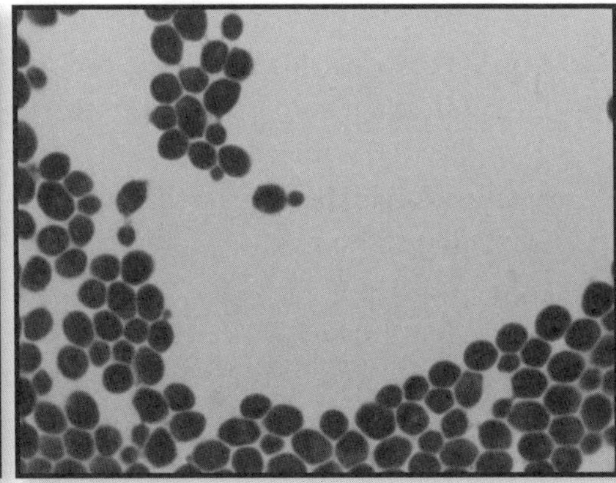

Saccharomyces cerevisiae

In Biblical times, wheat was generally used in bread making, but barley was a substitute among the poor (2 Kings 4:42). The first mention of bread in the Bible is Gen. 18:6 where Abraham asked Sarah to make cakes upon the hearth. Bread was also mentioned early on in Ex. 12:34 and Jer. 7:18. Bread was baked daily, no more than required for family use. Flour was mixed with water and made into dough and then rolled out into cakes. It was placed on the earthen floor, previously heated by a fire. The cakes were laid on the ground, covered with hot embers and baked, and then eaten. Also bread was put in troughs, bowls or flat plates and baked in portable ovens of earth, or upon heated stones or on the coals.

During the nomadic period, the Hebrews did not add leaven before kneading (Gen. 19:3, Ex. 12:8). They learned to leaven their bread while in Egypt (Ex. 12:15, Ex. 12:18-20). As they were departing Egypt, they took their dough before it was leavened, wrapped their kneading-troughs in their cloaks, and slung them on their shoulder (Ex. 12:34). Since they lived in Egypt 430 years, they learned to use the utensils of the country such as the kneading trough, which was a bowl of wicker or rush-work. Unleavened bread was also used in the Passover (Ex. 12:15-20, Ezek. 45:21) and the trespass offering (Lev. 6:17, 7:12). In Lev. 7:13 the first mention of leavened bread being used was when the Priest used it in the peace offering (Lev. 7:13). In the feast of Pentecost, the Hebrews were to bring from their homes two loaves as a special gift, that shall contain two tenths deals of flour and shall be baked with leaven (Lev. 23:17). The Law specified that bread baked with leaven should be used as save loaves offered to the Lord. These fine loaves were from wheaten flour, the quantity contained in them somewhat more than 10 pounds. As the wave-sheaf gave the signal for commencement, the two loaves symbolized the termination of the harvest season. They were the first fruits of the season being offered unto God by the priest in name of the whole nation (Ex. 34:22).

The loaves used at the Passover were unleavened; those presented at Pentecost were leavened. It is speculated that the difference is that one was a memorial of the bread hastily prepared at their departure, while the other was a tribute of gratitude to the Lord, for their daily food, which was leavened. In Lev. 24:1-23 the Israelites are commanded by God through Moses how to prepare the altar for a burnt offering, which is a repetition of Ex. 27:20-21. "Then you shall take fine flour, and bake twelve cakes with it…" (for the show-bread). These cakes were baked by the Levites, the flour being furnished by the people (1 Chron. 9:32: 23:29). Every Sabbath a fresh supply was furnished, but loaves were placed on the latter instead of the stale ones, which having lain for a week were removed and eaten only by the priests, except in cases of necessity (1 Sam. 21:3-6, Luke 6:3-4).

In making leaven bread, the Hebrews retained a small piece of the previous day's dough that contained the yeast and crumbled it into the water before mixing it with flour. Since this contains a high number of yeast, only a little is needed to leaven the bread (1 Cor. 5:6; Gal. 5:9). The Jews had to dedicate all food to God. An example of how this was done can be observed in Num. 15:20-21 where the first pinch of dough is offered to God and thus, the whole lump was made Holy. In Romans 11:16 Paul perhaps looked on Abraham as that first morsel of dough and his descendants as the lumps.

Bread is the most widely eaten food in the world. It provides a larger share of people's energy and protein than any other food and is often called the "staff of life". The origin of this phrase probably came from Lev. 26:26 where it mentions staff of bread. The Hebrew word Hehew, meaning bread, is often translated food. Bread was one of the main parts of a person's meal (1 Sam. 16:20) and probably consumed three loaves in one day (Luke 11:5-6); however, a prisoner like Jeremiah was allowed only one loaf per day (Jer. 37:21). In Jerusalem, Jeremiah was imprisoned on baker's street (Jer. 37:21), where there were probably a number of small bakeries. Withholding bread from the hungry was considered a sin (Job 22:7).

Bread was given as a gift (1 Sam. 9:7, 2 Sam. 16:1). Giving bread to the hungry was considered benevolent (Isa. 58:7, Ezek. 18:7). A special blessing was to be recited before eating bread (Ps. 14:4).

In the New Testament times, bread appeared like stones, which explains the temptation of Jesus by the Devil, who wanted Him to command stones to be made into bread (Matt. 4:3). "Man cannot live by bread alone..." (Matt. 4:4, Deut. 8:3). In John 6:48-58 Jesus states that He is the true bread of life. When the Hebrew people were wandering in the desert, the bread that they ate could not save them from death. Wealth, friends, food, etc. will not preserve life. There is need of something better than earthly blessings; there is need of that bread which cometh down from heaven, and which giveth life to the world.

Additional Reading:

Madigan, M.T., J.M. Martinko and J. Parker. *Brock, Biology of Microorganisms.* Pearson Education, Inc., NJ, 2003.

VINEGAR

Once wine was discovered, vinegar was not far off. The name vinegar is derived from the French word vinaigre, which means sour wine. It is prepared by allowing a wine to go sour under controlled conditions. But many kinds of vinegar are not made from wine. In Num. 6:3 it mentions "vinegar of strong drink," which might have been made from palm juice.

This sour liquid results from two types of biochemical changes: (1) an alcoholic yeast fermentation of a carbohydrate and (2) oxidation of the alcohol to acetic acid by vinegar bacteria.

A yeast such as *Sacchromyces cerevisiae* can be used for the production of alcohol. Then the acetic acid bacteria species of the genus *Acetobacter,* oxidizes the ethyl alcohol to acetic acid. These are alcohol-loving strains of bacteria, *Gluconobacter* and *Acetobacter,* acetic acid bacteria that oxidize ethanol, which turn sweet wine into sour vinegar. There are controlled industrial processes now available for its production, which is not left to so much chance as it was in Biblical times.

The acid nature of vinegar is suggested in Prov. 10:26, 25:20. Psalms 69:21 indicates it was used in punishment. In Prov. 25:20 it suggests that vinegar added to niter cause effervescence.

The vinegar used by Ruth and Boaz in Ruth 2:14 was typical of that used in wine-growing countries. It was particularly used to enhance the flavor of food. The posca of the Romans was similar in nature and formed part of the soldiers' rations. Posca was offered to Christ on the cross as refreshment (Mark 15:36; John 19:29-30). This was different from the myrrh-flavored vinegar considered medicine to relieve pain, which Jesus had refused earlier (Matt. 27:34; Mark 15:23) in fulfillment of Ps. 69:21. The use of vinegar as medicine is also mentioned in Prov. 25:20, where it is used to dress a wound. Vinegar was prohibited to Nazarites (Num. 6:3).

As a final thought, for more indepth knowledge of microbiology the main organisms converting alcohol to acetic acid is *Acetobacter aceti.* The acetic acid bacteria comprise an ecological group, consisting of Gram-negative, rod-shaped bacteria, which grow characteristically in plant materials that have undergone an alcoholic fermentation. These bacteria convert the alcohol to acetic acid. The acetic acid bacteria are subdivided into two genera: the polarly flagellated *Gluconobacter,* where the organs of locomotion are inserted at the pole or poles and the peritrichously flagellated *Acetobacter,* where the flagella are inserted at many points along the sides of the cell.

Since the Gram stain is first mentioned in this chapter, it should be described further because of its great value in the identification of bacteria. A film of the bacteria is made on a slide and is stained with the dye crystal violet. It is taken up by bacterial cells indiscriminately. Next, iodine is added, which enters the cells and forms large complexes with the crystal violet. Then a decolorizing agent alcohol is added, which removes the complexes from some types of organisms but not from others. Last, a red dye, safranin, is applied. This procedure dif-

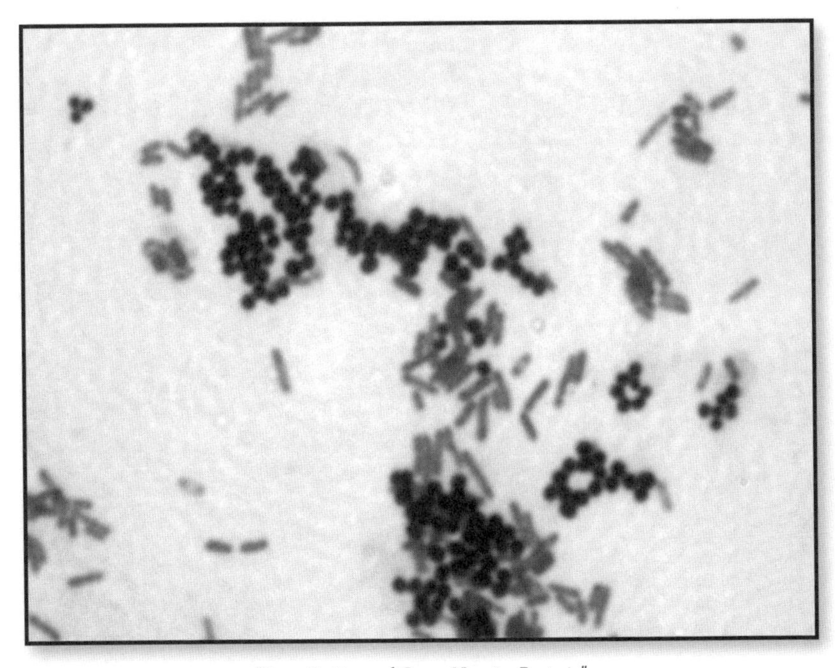

"Gram-Positive and Gram- Negative Bacteria"

ferentiates bacteria into those that retain the first stain—the purple colored gram-positive bacteria and those that have been decolorized show the second stain and are red gram-negative when viewed under the bright-field microscope. This differential staining procedure is very important in identifying bacteria to know whether it is gram-positive or gram-negative. This staining method, devised in 1884 by the Danish physician, H.C. Gram, represents an important taxonomic feature that is dependent on the cell wall structure and also correlates with other characteristics such as susceptibility to antibiotics.

Additional Reading:

http://www.keyway.ca/htm2001/20011228.htm

OLIVES

For millennia olives have been used as sources of oil and food. The olive is a fruit that grows in regions near the tropics. It is first mentioned in Gen. 8:11, when the dove returned to the ark with an olive leaf to Noah after the flood. Olive trees were common in the land of Canaan (Ex. 23:11; Deut. 6:11; 8:8). Cultivation of the olive tree was one of the most important of the farmer's activities (Deut. 28:40). It was deemed worthy of being called the king of the trees (Judges 9:8). Solomon used olive wood for some of the doors inside the original Temple (1 Kings 6:31-33). Prophets had even described Israel as a beautiful, fruitful olive tree (Jer. 11:16; Hos. 14:6). Branches of the

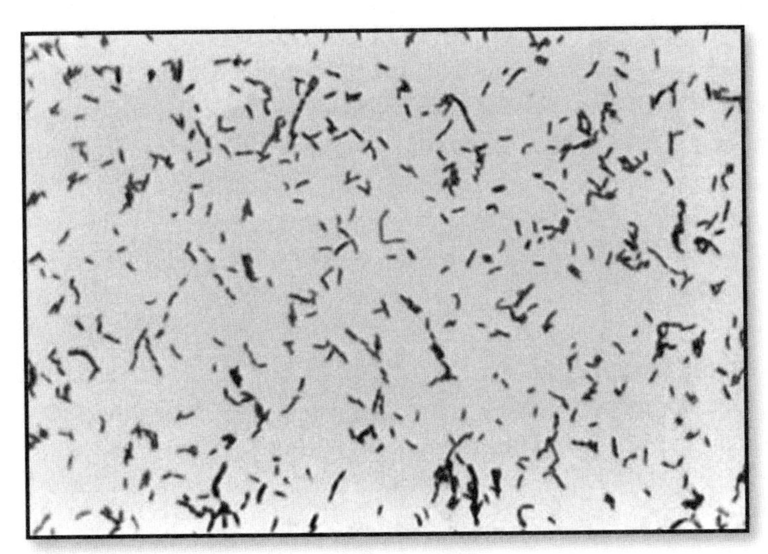

Lactobacillus mesenteroides

olive tree were used for booths (huts) (Neh. 8:15) and were noted to bear flowers (Job 15:33). Paul used an illustration concerning the olive tree to explain the purpose of God and how it involved the Jewish rejection of Christ and the Gentile response to him (Rom. 11:17-24).

The olives ripened in autumn and were harvested toward the end of November. It was harvested as described in Deut. 24:20 and Isaiah 17:6, 24:13, by shaking or beating with poles. The most important material from the olive is the oil. The oil was usually extracted from the berries by placing them in a shallow rock cistern and crushing them with a large upright millstone. Sometimes the fruit was pounded by the feet of harvesters (Deut. 33:24; Micah 6:15). Olive oil fed wicks for illumination (Matt. 25:3) and was used as a medium of exchange. The oil was used to fuel the Menorah in the Tabernacle in the wilderness (Lev. 24:1-2), and later in the Temple in Jerusalem. It was also a base for ointments and hair tonics and used in medicines (Luke 10:34) and as a food (2 Chron. 2:10).

Fresh or pickled olives eaten with bread formed an important part of ancient Palestinian diets. The Romans apparently pickled olives by placing them in a solution composed of 1/3 brine, 1/3 vinegar, and 1/3 boiled grape juice. This mixture was placed in vessels that were sealed with pitch. The olives were not pleasant to the taste and they chopped the olives and mixed them with rue, parsley and mint. The use of lye to hydrolyze the bitter glycoside, oleuropein, appears to be a modern invention. Olives to be pickled are done so by the natural flora of green olives, which consists of a variety of bacteria, yeast, and molds. During the brining, now done in oak barrels, the lactic acid bacteria become prominent during the intermediate stage of fermentation. *Leuconostoc mesenteroides* and *Pediococcus cerevisiae* are the first lactics to become prominent, and then are followed by the more aciduric lactobacilli with *Lactobacillus plantarum* and *brevis* being the most important. The fermentation may take as long as 6-10 months, and the final product has a pH of about four following up to 1% lactic acid produced by the bacteria. Various factors, such as the origin, maturity and variety of olives, and treatment prior to brining, brine strength, sugar content, acidity, available desired microflora and temperature influence the fermentation or pickling process. *L. mesenteroides* are gram-positive cocci in chains, while *Pediococcus* are gram-positve cocci in pairs, singly, short chains, or in tetrads. The lactobacilli are gram-positive long, slender, rods.

Although the Bible does not state that they fermented olives and ate them, the Romans knew the technology so the Hebrews were exposed to pickled olives. Even today it is the characteristic tree of Palestine yielding olives that produce highly valued oil. We should all be familiar with Psalms 52:10, "I am like a green olive in the House of God; I trust in the love of God for ever and ever."

Additional Reading:

http://web.odv.edu/webroot/instr/sci/plant.nsf/pages/olive

Frazier, W.C., and D.C. Westhoff. *Food Microbiology*. McGraw-Hill Book Co., 1988.

Cheese & Butter

Cheese is a healthful food made from milk. The first cheese was probably made over 4,000 years ago by Nomadic tribes in Asia. This probably started when they first took milk from wild or domestic animals. Unless the milk was used fairly quickly, it became sour and formed an acid curd by natural means and released whey, leaving a semisolid curd. This was the first kind of cheese and this method of production is still used by primitive people.

The manufacture of all cheeses depends upon the activity of microorganisms. There are now over 400 kinds of cheese and most of them could be made from the same batch of milk. This is possible by providing conditions that favor the development of selected types of microorganisms.

The initial curdling of milk was probably caused by *Lactococcus lactis,* which is a small gram-positive coccus in pairs and chains. The organism ferments the lactose in milk with the principle end product being lactic acid. One of the other main acid-producing bacteria found in milk is lactobacilli, which are long slender gram-positive rod-shaped (non-motile) bacteria. Some representative species that ferment lactose to lactic acid are *Lactobacillus casei, L. acidophilus, L. lactis* subspecies *lactis* and *L. bulgaricus.*

Lactococcus lactis

After the milk has curdled, the curd is separated from the

whey. The type of cheese to be produced determines how much moisture must be removed. For some purposes, the whey is permitted to drain off. The resulting curd is soft and has high moisture content. Low moisture cheeses are obtained by heat, pressure, and cutting the curd. After the moisture content has been reduced to the desired limit, the curd is pressed into a characteristic size and shape. Salt is added to almost all cheeses. It may be mixed with the drained curd, or it may be applied to the surface of the pressed form. Salting enhances flavor, controls moisture by withdrawal of water, and controls development of spoilage microorganisms.

Some types of cheese consist of fresh curd like cottage cheese. Mixing cream into this type of curd produces creamed cheese; these are unripened cheeses. Most cheeses require ripening by bacteria or molds after the curd is pressed into form. Cheese ripening involves the growth of mixed populations of microorganisms in and on the complex substrate. To distinguish all the biochemical events that take place and to attribute each of these changes to a specific organism is difficult and beyond the scope of this book.

In Biblical times cheese was produced by salting the strained curds, molding them into small disks about the size of a hand and drying them in open air (Job 10:10). There were several varieties of cheese: hem'ah or cream (Prov. 30:33), gebinah or ordinary cheese (Job 10:10), and hariz he-halab (1 Sam. 17:18), probably like cottage cheese.

Twice in the Old Testament cheese was included in the list of provisions that David took to his brothers (1Sam. 17:17-18), and it was among the provisions that David received at Mahanaim (2 Sam. 17:29). In Moses' song concerning the perfection of God cheese is mentioned (Deut. 32:14). Job must have had knowledge of cheese since in Job 10:10 he says, "didst thou not pour me out like milk and curdle me like cheese."

The Bible mentions goat's milk and camel's milk. Both were prized for drinking as well as for making cheese and butter. Butter is mentioned eleven times in the Old Testament. Biblical butter is known as samneh, which is a clarified butter that can be kept for long periods of time without turning rancid. In Isa. 7:14-15, it is prophesized that a virgin shall conceive and bear a son, and His name shall be called Emmanuel. Butter and honey shall He eat. Both cheese and butter making are ancient processes that rely on fermentation of the milk sugar lactose to form lactic acid by *Lactobacillus.* Butter-making typically utilizes *Lactobacillus lactis* subspecies *cremoris,* where cream is used at the starting material and is made by churning (Prov. 30:33). The butterfat globules coalesce into granules. The liquid portion, buttermilk is drained off, and the granules are further processed. The cream from which the butter is churned would contain some of the same bacteria discussed in connection with cheese. The Romans used butter as a hair dressing and as skin cream. Abraham entertained three angels by feeding them milk, butter, and a calf that he had dressed (Gen. 18:8). In Judges 5:25 Deborah and Barak speak of butter in a lordly dish. Also among the provisions that David received at Mahdnaim was butter (2 Sam. 17:29). Zophar mentions butter in Job 20:17. Buttermilk, as the name suggests, is the milk that remains after cream is churned for the production of butter. But when Sisera thirsted, "Milk she gave him in a lordly bowl" she brought him buttermilk since one does not drink butter (Judges 5:25). Bedouins prepare sour milk, or leben, by pouring the fresh milk into a skin. In Judges 4:19 it states, "She opened the milk-skin (a bottle of milk), and gave him drink. The milk clots by microbial action on the sides of the skin. When the process of fermentation begins the skin is shaken vigorously and the milk is served with refreshing sourness.

As we can see, over the centuries, fermented milk production assumed a key place in the diet of the Hebrews.

Additional Reading:

http://genome.jgi-psh.org/draft~microbes/lacca/lacca.home.html

Carr, F.J., D. Chill, and N. Maida. *The Lactic Acid Bacteria: A Literature Survey.* Crit. Rev. Microbiol. 28(4):2810370, 2002.

Pickles

The history of pickles dates back 4,000 years ago to Mesopotamia, where it is believed cucumbers were first preserved by fermentation. Cucumbers native to India were brought to the Mesopotamia area. Cucumbers were found in Egypt as recorded in Numbers 11:5, probably the species *Cucumis chate.* Isaiah speaks of a garden of cucumbers in Isa. 1:8. Many famous people liked pickles, including Cleopatra and Julius Caesar. In 1820, Frenchman Nicholos Appert, the father of canning, was the first person to commercially pack them in jars. Christopher Columbus brought pickles to the New World. George Washington, John Adams and Dolly Madison liked the taste of pickles. In colonial America, the pickle patch was important to their livelihood. During my grandparent's era, they fermented cucumbers in stone crocks, and canned them in glass jars.

Pickles start as cucumber seeds. I typically plant about four to five seeds to a hill and with a few hills more cucumbers are produced than you can eat. To preserve the extra cucumbers you make pickles. Young cucumbers are placed in brine with various spices and flavorings and covered to prevent access of oxygen. If allowed to stand for several days to a week, various bacteria normally present on cucumbers begin to grow. First, varieties of *Enterococcus* and *Entrobacter* would grow followed by *Leuconostsoc mesenteroides, Pediococcus cerevisiae, Lactobacillus brevis* and *Lactobacillus plantarum.* Of these the Pediococci, and *L. ploantarum* are the most involved, with *L. brevis* being undesirable because of its ability to produce gas. *L. plantarum* is the most essential species in pickle production. Salt is increased from 5% to 16% during the six to nine week fermentation. The salt inhibits undesirable gram-negative bacteria and extracts water and water-soluble constituents from the cucumbers such as sugars, which are converted by the above lactic acid bacteria to lactic acid. The product is a salt-stock pickle from which pickles such as sour, mixed sour, etc. may be made.

To leave chance out of the succession of microbes just described, picklers sometimes add vinegar to the brine to drop the pH significantly. This acetic acid immediately inhibits the growth of most spoilage bacteria and gives a firmer product. This does not shorten the time of fermentation significantly.

There are over fifty different types of pickles in grocery stores today, with dill leading in popularity. Since anise or dill, *Anethum graveolens* is mentioned in the Bible it is possible that the Hebrews made dill pickles. In Matt. 23:23 it is mentioned that this plant, its stem, leaves and seed, was subject to tithe. This plant is an annual or biennial herb growing wild in lands bordering on the Mediterranean. The seeds can be used as a carminative in medicine in the form of "dill water". Dried powdered seeds of the Coriandor (Chinese Parsley) mentioned in Ex. 16:31 and Num 11:7 were used to season pickles and meat.

Additional Reading:

http://www.fao.org/docrep/x0560e10.htm

FLAX

Flax (*Linum usistatissimum*) has been grown through out the world for thousands of years for its beneficial uses. The use of linen, or "flaxen cloth", dates back to people who lived about 10,000 years ago. These people dressed in skins but they made coarse cloth and fish nets from flax. Fragments of this cloth and nets have been discovered in parts of Switzerland, the home of the Neolithic Lake Dwellers. Fine linens have been the burial shrouds of the Egyptian Pharaohs (Deut. 22:11) and the textile of Biblical times (Deut. 22:11).

It is interesting that microbes were used to benefit society well before people realized that they were taking advantage of these minute microbes. There are two general types of flax: one is grown for seeds, the other for its fiber. The oil of the seeds is the well-known linseed oil. The flax plant produces a stem growing thirty to forty-seven inches, with pretty blue flowers. Flax was early cultivated in Palestine (Prov. 31:13, Josh. 2:6), and was boiled (Ex 9:31) when the seed vessels reached maturity and the plant was ready for gathering. The failure of the flax was one of God's judgments (Hos. 2:9). The stalks were dried on the housetop (Josh. 2:1), then soaked in water and then the fiber combed out (Isa. 19:9). The tow (wick) of Isa. 43:17 is teased out of flax.

In Biblical times there were two traditional methods used to ret flax for fiber: water-retting and dew-retting. The retting process begins by bundling flax plants and drying them in stacks. The object of retting is to loosen the flax fiber from the outside woody stalk. In water-retting, stacks of flax stems are submerged in rivers, lakes or placed in long trenches several feet deep, covered with water, and weighted down with stones to exclude as much air as possible. The anaerobic bacteria colonized the flax stems and degrade pectin and other matrix compounds freeing fibers from the core tissue. The enzyme pectinase is particularly from *Clostridium* species, and specifically from *Clostridium pasteurianum,* a gram-positive rod-shaped bacterium. After a few days, the water turns black and an unmistakable stench signaled that retting or rotting was taking place. After about two weeks the flax is soft and pliable so that the fibers could be easily removed by pounding with wooden blocks.

Dew retted flax is spread on fields in thin layers and allowed to weather in the dew and rain. Dew-retting depends mostly upon plant cell-wall degrading enzymes produced by indigenous, aerobic fungal consortia. Fungi are eukaryotic microbes, which will be described in more detail in a later chapter. In dew-retting, flax plants are pulled from the soil and laid out in fields for selective attack by fungi which can take up to six weeks. The flax is turned over three to four times during this time to ensure even retting.

It could be argued that flax was the most important plant fiber in Biblical times since it was used to make linen. Living up to its species name, *usitatissimum* means useful, the flax plant, used as a source of fiber for the making of blankets, handkerchiefs, paper, clothing and sails, became a valuable commodity during the 17th and 18th centuries. Maybe America would not have been discovered without the use of the flax plant. The use of flax/linen as an acceptable sail material was especially important to the world's great explorers like Christopher Columbus. The renowned Mayflower, along with many other ships associated with America's early history, had sails made from this plant's fibers.

For numerous citations concerning flax and linen in the Bible visit the first website listed below. For a description of a modern process using water retting for the production of linen, read about the technology that is now being used in Iceland.

Additional Reading:

http://web.idu.edu/webroot/instr/sci/plant.nsf/pages/flax
http://www.randburg.com//is/flax/

Coral

The word "coral" is derived from the ancient Greek word "korallion", which refers to the red coral of the *Mediterranean* Sea, *Corallium.* Currently there are twenty-five species of *Corallium* recognized. The one that is most likely collected today that is used for jewelry and beads is Corallium *rubrum.* Coral is a limestone formation produced as a calcareous exoskeleton around the polyps. The animals that form coral belong to the same animal group as the hydras, jellyfish, and sea anemones. Most individual coral animals, called polyps, are less than one-inch in diameter but a small percentage of them measure one foot. A coral polyp has a cylindrical shaped body. At one end is a mouth surrounded by tiny tentacles. The other end attaches to hard surfaces in the sea. When the animals die, they leave limestone "skeletons" that form the foundations of barriers and ridges in the sea called coral reefs.

Red Coral

Coral reefs appear as sea gardens, since many colorful sea animals live among the corals. Sometimes coral masses build up until they rise above the water to form coral islands.

Most coral polyps live together in colonies. The stony corals attach themselves to each other with a flat sheet of tissues that connects to the middle of each body. Half of the coral polyp extends above the sheet and half below. Coral polyps reproduce either sexually from eggs or asexually by budding. Coral polyps build their calcium carbonate skeletons by taking calcium out of the seawater. There are two types of coral: hermatypic and ahermatypic. Ahermatypic corals usually do not form huge exoskeletons and individual polyps do not harbor symbiotic algae. Hermatypic corals are found in shallow tropical seas, producing massive exoskeletons and are characterized by individual polyps containing symbiotic algae within the gastrodermal cells. These one-celled algae living in the internal tissue are called zooxanthellae. This symbiotic relationship allows for the production of enough calcium carbonate for coral reefs to originate and grow. Corals that take part in symbiosis are called zooxanthellate corals. The coral provides a protected habitat for the algal cells but utilizes products of algal photosynthesis to produce more calcium carbonate than it could without the algae. A well-known algae in this symbiotic relationship with corals are the dinoflagellates. These photosynthetic, unicellular algae will be mentioned later when the plagues are mentioned.

Hebrews used coral for beads and ornaments. It ranked among precious stones (Job 28:18; Ezekiel 27:16). Most precious stones mentioned in the Bible are: the stones in the high priest's breastplate (Ex. 28:17-20; 39:10-13), those in the covering of the King of Tyre (Ezek. 28:13), and those in the foundations of the New Jerusalem (Rev. 21:19-20).

The coral flora would not be the same as in Biblical times because destruction of coral is and has occurred because of tourism, rise in seal level, increase of sea temperature, pollution and the predatory starfish acanthaster. It is difficult to know whether the coral harvested by the Hebrews was made into jewels, but possibly coral is a rendering of the Hebrew word peninim, which is ruby (Prov. 3:15; 8:11; 20:15 and 31:10). They could have obtained red corals from the Mediterranean Sea themselves or by trade. Red corals can be orange, pink, to white.

They are now commonly used in beads, jewelry and carvings. Red coral of the species *Corallium rubrum* or *Corallium japonicum* appears to have been found in ancient Egyptian graves.

For general information on corals visit the web site below.

Additional Reading:

http://www.coralreefnetwork.com/educate/shows/foodwebs/slide5.htm

Spoilage of Grains & Grain Products

Mildew. Mildew can be considered decay produced in living and dead vegetable matter, and in some manufactured products of vegetable matter by parasitical fungi. Fungi are multicellular filamentous, usually nonmotile organisms that have no true stem, roots, or leaves and lack chlorophyll. Molds, mildews, yeasts, mushrooms, and puffballs are common fungi. There are two main classes that damage plants: powdery mildew and downy mildew. One of the most serious powdery mildews is *Unicinula necator,* the cause of powdery mildew of the grape vine, which under favorable conditions to the fungus can result in the complete destruction of the entire crop in a region.

There are about fifty different kinds of powdery mildews, and some can attack several different plants. Approximately 1,500 different kinds of flowering plants may be infected by powdery mildew. Over 2,000 species of Palestinian flora are known to exist. The microbes generally grow on the outside of stems and leaves and the spores formed produce a powdery coating that can be seen with the unaided eye.

Downy mildews produce yellow spots on the upper surfaces of the leaves or young fruits. Economically they are among the most important fungi. Their hosts include a large number of plants. A destructive representative is the downy mildew of grapes caused by *Plasmopara viticola.*

Other serious plant pathogens are the basidiomycete species; the most important are rusts and smuts. Rusts and smuts can attack hosts so severely that they kill the plant. In economic terms, rusts that attack cereals such as barley, corn, oats, rye, and wheat are most important, causing losses of millions of tons of grain annually. Grasses of various kinds are the commonest victims of smut, and destruction of the grain or fruit in which humans are interested gives these fungi great economic importance. Rusts attack the tissue of a wide variety of seed plants and ferns. During the life history of both smuts and rusts, several different types of spores may be produced. Ordinarily the sexually produced basidiospores are those that bridge seasons of prolonged cold or drought.

In Deut. 28:22, Amos 4:9, and Hag. 2:17, mildew was regarded as God's punishment on the disobedient. In the preceding references along with I Kings 8:37 and 2 Chron. 6:28, the Bible associates mildew with blasting or scorching, which was a drying of plants by hot wind from the south. In the latter two verses Solomon prayed for deliverance from mildew. In damp situations, species of mildew grow on the walls of the Hebrew homes, which caused illness and instability to their house (Lev. 14:54-57).

The clothing can also be contaminated by mildew. In Lev. 13:47-50 it states that if any clothing is contaminated with mildew, any woolen or linen clothing, any woven or knitted material of linen or wool, any leather or anything made of leather, if the contamination in the clothing or leather, or woven or knitted material, or any leather article, is greenish or reddish, it is spreading mildew and must be shown to the priest. The priest is to examine the mildew and isolate the affected article(s) for seven days.

It is difficult to speculate specifically which type of mildew the Bible discusses, since the scripture mentions only mildew. Certainly the mildews were a problem in Biblical times.

Molds. Besides fungi causing serious plant diseases, some of the molds cause spoilage of food. Molds are filamentous fungi that grow in forms of a tangled mass. The totality of the mass is referred to as mycelium. Mycelium is composed of branches or filaments termed hyphae. *Rhizopus* hyphae are nonseptate, which means they do not have crosswalls. In Joshua 9:5, 12, the bread is said to be moldy. The bread used by travelers was like a biscuit, made in the form of large rings, about an inch thick, to four or five inches in diameter. Not being as well baked as modern biscuits, it becomes hard and moldy from

Rhizopus oryzae

the moisture left in the dough. It is usually soaked in water prior to being used. The common bread mold *Rhizopus* forms a cottony, white growth while their spores are black thus they are considered a black mold. The most commonly encountered species is *Rhizopus stolonifer.* Not only does it spoil bread but produces watery soft rot of apples, pears, stone fruits, grapes, figs, and others. Some cause "black spot" of beef and frozen mutton.

Additional Reading:

http://www.apsnet.org/online/feature/pmildew/gallery.htm

http://www.inra.fr/internet/produits/HYP3/pathogene/6plavit.htm

http://www.biologie-uni-hamburg.de/b-online/e33/33c.htm

www.extento.hawaii.edu/kbas/crop/type/rsfolo.htm

DISEASES

The term disease appears to have originated from Latin stems that mean "living apart," a reference to the separation of ill individuals from the general population. Disease is any change from the general state of good health. In this chapter we will specifically deal with the infectious disease process, which are thought to consist of three stages: (1) microbial entry and colonization of the host; (2) microbial invasion and growth in host tissues along with elaboration of toxic substances; and (3) the host response. These stages reflect the concept of infection, which is the presence of microbes in the host. The disease occurs when an infection results in any change from a state of health.

Among primitive races, disease was either regarded as the result of hostile magic putting a spell upon a person, or else its incidence was ascribed to the violation of a taboo. In Old Testament times, disease was believed to be a punishment sent from God for human wrong-doings (Ex. 4:11; Deut. 32:39). Another view ascribed to the origin of disease was the work of the Devil (Job 2:7). In New Testament times diseases were called spirits of dumbness (Mark 9:17; 9:25). Also, disease was thought to occur because of some sin the person had committed (John 9:2).

An early concept of disease was by the Italian physician Fracastorius (1485-1553), who postulated that disease was transmitted by invisible particles or seeds from one person to another or from contact with clothing or utensils of the infected. Two hundred years elapsed before Anton van Leeuwenhoek made a description of microbes in 1685. The theoretical explanation of infectious disease as proposed by Fracastorius was not supported by experimental proof until the Koch-Pasteur era (1870-1890), particularly Robert Koch through the sequence of isolation, reinfection, and recovery of the infective agent developed Koch's postulates by which it is possible to establish the causative agent of infectious disease.

Thus, obviously Bible references concerning specific diseases are general and vague. Even where concrete mentions of particular diseases are made, it is difficult to decide what the exact nature of the sickness was. Sometimes the symptoms are given, though sometimes very indefinitely. So with this information I will discuss with the best of my knowledge microbial diseases that are presented in the Bible. I will do this in alphabetical order except where grouping of topics makes for more logical presentation.

Anthrax. According to Ccieslak and Eitzen in 1999, the fifth and sixth plagues in Ex. 9:1-12 may have been outbreaks of anthrax in cattle and humans, respectively. God used biological warfare on Pharaoh's kingdom, while the Israelites and their livestock were immunized. In Ex. 8:13 it states the frogs died possibly because they had been infected by *Bacillus anthracis* from the Nile algae. Possibly the flies of the fourth plague carried the anthrax bacilli that would now infect the animals.

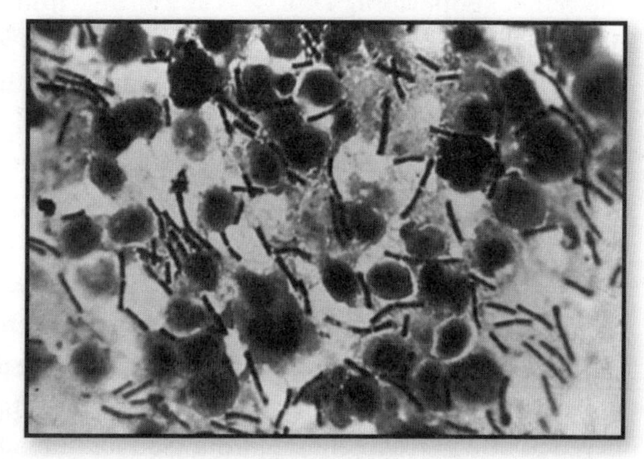

Bacillus anthracis

Anthrax is a zoonosis, which infects animals that can be transmitted to humans by direct contact with infected animals such as cattle, goats, sheep or their products. The *Bacillus anthracis* endospores are ingested along with grasses, causing a fulminating, fatal septicemia. Meat-eating animals can acquire the disease by eating infected flesh or by inhaling anthrax endospores. The endospore-forming bacillus is a large, aerobic, gram-positive bacterium that is apparently able to grow slowly in soil types having specific moisture conditions. The endospores have survived in soil tests for up to 60 years.

People at risk for anthrax are those who handle animals, hides, wool, and other animal products from certain foreign countries. Goat hair and handicrafts containing animal hides from the Middle East have been a repeated source of infection. Pulmonary anthrax, or "wool sorters' disease", was such a problem in nineteenth-century England that legislation was enacted to protect textile workers from this occupational disease. Human infection is usually through a cut or abrasion of the skin, resulting in cutaneous anthrax. If endospores reach the intestine, gastrointestinal anthrax may result.

In humans the incubation period for cutaneous anthrax is one to fifteen days after endospores enter epithelial layers of the skin. Lesions called eschar develop at the site of entry one to three cenimeters in diameter. The center of the lesion becomes black and necrotic. Headache, fever, and nausea are the major symptoms. Pulmonary anthrax resembles influenza. If the bacteria invade the bloodstream, the disease can be fatal. Symptoms of intestinal anthrax closely mimic those of food intoxication. The signs and symptoms of anthrax are due to anthrax toxins, a complex exotoxin system composed of three proteins. *B. anthracis* not only has unusual toxins it also has an unusual capsule composed of d-glutamic acid.

Of course, bioterrorists operate outside the laws and morals of society and on September 11, 2001, the United States experienced the worst terrorist attack in its history. *Bacillus anthracis* endospores had been intentionally distributed through the postal system, causing twenty-two cases of anthrax, including five deaths. These terrorists miss the total teachings of God as revealed in the Holy Bible concerning loving your fellow man.

Blindness. Blindness is exceedingly common among the nations of Palestine. The words describing this affliction are of frequent occurrence in the Bible, sometimes in the literal, sometimes in the metaphorical, sense. Apparently only two forms of blindness are recognized: (1) a highly infectious disease caused by microbes, aggravated by sand, sun-glare, and dirt which often renders the eyes useless, and; (2) that due to old age.

The disease of eyes were common in biblical times because of the spreading of the microorganisms of eye disease by close contact, filth, and flies. Leah, Labon's daughter, had sore eyes (Gen. 29:17) apparently caused by trachoma, which is a severe conjunctivitis with complications such as permanent scarring. Trachoma is currently the leading infectious cause of blindness and preventable blindness globally. It is caused by specific strains of *Chlamydia trachomatis.* The bacterium is a small gram-negative rod that grows as an obligate intracellular parasite. The disease begins abruptly with an inflamed conjunctiva. This leads to an inflammatory cell exudate and necrotic eyelash follicles beneath the conjunctival surface. Blindness is a result of long-term mechanical abrasion of the cornea by scars and by turned-in eyelashes. Secondary infections by other bacterial pathogens are also a factor.

Ague or malaria mentioned in Lev. 26:16 can cause blindness.

Boils and Blains. Boils or furuncles are an inflammation of tissue, a localized pyogenic infection originating in a hair follicle. Blains are sores on the skin, blisters or blotches, thus different microbes can cause these conditions. Our skin accounts for fifteen percent of our body weight. It provides an effective barrier to invasion by most microbes except when it is damaged.

As mentioned in Ex. 9:1-12 the 6th plague might have been boils on humans caused by the anthrax bacillus. However, a more common cause of boils is *Staphylococcus aureus. S. aureus* is a common cause of skin conditions, usually appearing as localized lesions such as boils, impetigo, and sties. An infection of the base of an eyelash is called a sty. Impetigo is a highly contagious pyoderma caused by staphylococci, streptococci, or both. Pyoderma is a pus-forming skin infection, caused by staphylococci, streptococci, and corynebacterium, singly or in combination.

Staphylococci are spherical gram-positive bacteria that form irregular clusters like grapes. For most clinical purposes, these bacteria can be divided into those that produce coagualse, an enzyme that coagulates (clots) fibrin in blood and those that do not (coagulase-positive and coagulase-negative strains respectively).

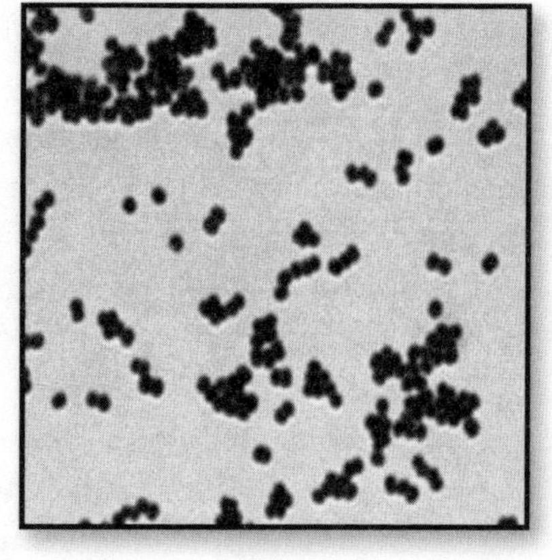

Staphylococci

S. aureus is the most pathogenic of the staphylococci. Many times, it forms golden-yellow colonies on laboratory media, and is coagulase-positive. There is a high correlation between the bacterium's ability to form coagulase and its production of damaging toxins, several of which many injure tissue.

The nasal passages provide an especially favorable environment for *S. aureus,* which are often present there in large numbers. Its presence on unbroken skin is often the result of transport from the nasal passage. Staphylococci are the primary cause of a troublesome problem in hospital nurseries, impetigo of the newborn. Symptoms of this disease are thin-walled vesicles on the skin that rupture and later crust over. *Streptococcus pyogenes* that will be discussed in more detail later can also cause impetigo. This is the most common in toddlers and children of grade-school age. Streptococcal impetigo is characterized by isolated pustules that become crusted and rupture. The disease is spread mainly by contact, and the microbes penetrate through minor abrasions or insect bites.

In Isa. 1:6 the Bible describes wounds with pus (abscess). In Lev. 13:28 the Bible describes an infection of the skin without pus.

In 2 Kings 20:7 Isaiah treats a boil by the use of figs. This treatment is also described in Isa. 38:21.

In Job 2:7 boils were probably not boils since they rarely cover the whole body. This skin condition might have been leprosy, syphilis or ringworm.

In Lev. 13:18 blains might have been bright spots as mentioned in Lev. 13:2. These general terms might mean freckles, or some inflammation as boils, leprosy, measles or scarlet fever.

The Lord gives regulations about infectious skin diseases such as boils in Lev. 13, 14, and 15.

Brucellosis (Undulant Fever). While Paul was voyaging toward Rome his ship wrecked and the survivors ended up on the island of Malta. Publius, the governor on the island, generously hosted Paul and the others for three days (Acts 28:1-10). Paul was able to return the favor by healing the governor's father. He laid his hands on his host's sick father and healed him. The ailing man might have been suffering from "Malta fever", an illness traced to a microbe in the milk of Maltese goats and a common illness on the island during this time.

Brucellosis also called Malta fever, is a zoonosis highly infective for humans. A zoonosis is a disease that can be transmitted from animals to humans. Brucellosis is an infection caused by bacteria of the genera *Brucella.* Human infection results from occupational contact with infected animals or from ingestion of infected milk, milk products, or tissues. The symptoms of brucella are often non-specific and include fever, malaise, and weight loss, often without physical findings.

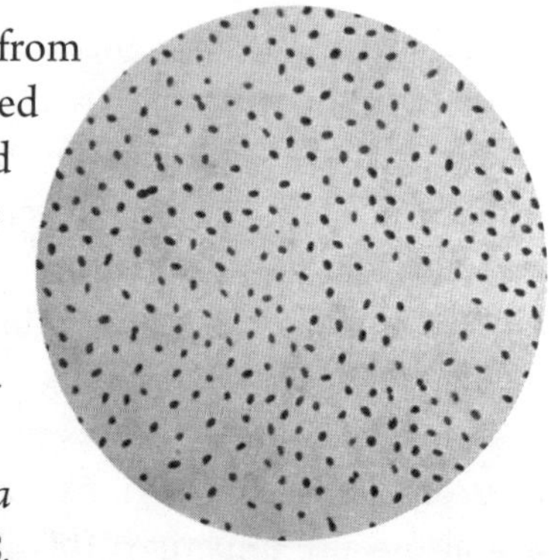

Brucella melitersis, initially named Micro - coccus melitensis. Photo by CDC.

There are four species of *Brucella* that cause infection in humans. *B. melitensis* was isolated in 1887 on the Mediterranean isle of Malta by Sir David Bruce. The most pathogenic is *B. melitensis,* followed by *B. suis* in swine, *B. abortus* in cattle, and *B. canis in* dogs. Cross-species infection occurs (e.g., *B. abortus* in sheep) and other animals may become infected.

Brucella are small, nonmotile, nonencapsulated, gram-negative coccobacilli. After invading the body, they multiply inside phagocytic cells and move through the lymphatic system into the blood, where they cause an acute bacteremia, within one to six weeks. Bacteremia may result in foci in cells of the reticuloendothlial system in the liver, spleen, and bone marrow and in other organs such as the kidneys. The reaction of tissue to *Brucella* is the formation of granulomas. *Brucella* also tend to multiply in the uterus of a susceptible animal.

Brucellosis has a gradual onset with a daily fever cycle-high in the afternoon and low at night after profuse sweating. This fever pattern gives the disease its alternate name, undulant fever. These episodes are caused by the release of bacteria from granulomas into the bloodstream. The spleen, lymph nodes, and liver can be enlarged, and jaundice can be present. The initial acute phase lasts from several weeks to six months. Infections that last for more than twelve months can result in colonization of the bones, joints, kidney, spleen, or heart valves. When death does occur, it is usually from endocarditis.

Of course when the natives on the island saw the miracle of healing, the news spread rapidly. Soon people from all over the island were coming to Paul for healing. He healed many diseases of people on the island during his three-month stay. One could only imagine what these diseases were since these people had no hygienic rules, drugs, and vaccines were not available.

Bubonic Plague. Discussions of pestilence or plague are very common in the Old Testament. The bubonic plague was a widespread affliction of Bible lands. Few diseases have affected human history more dramatically than plague; known in the Middle Ages as the Black Death. During the Black Death, plague caused 20 to 30 million deaths in Europe. The disease is caused by a gram-negative, rod-shaped bacterium, *Yersinia pestis*. Normally a disease of rats, plague is transmitted from one rat to another by the bite from an infected rat flea. If the host dies, the flea sucks a replacement host, which may be another rodent or a human. A plague-infected flea is hungry for a meal because the growth of the bacteria blocks the flea's digestive tract, and the blood the flea infests is quickly regurgitated. From the fleabite, bacteria enter the human's bloodstream and multiply in the lymph and blood. The lymph nodes in the groin and armpit become enlarged, and fever develops. Such swellings, called buboes, account for the bubonic plague. Other symptoms might be, headache, extreme fatigue and muscle aches. There are two other types of plague septicemic and pneumonic. Symptoms of plague vary depending on the type and on how an individual contract's it. It is possible to develop more then one type of plague.

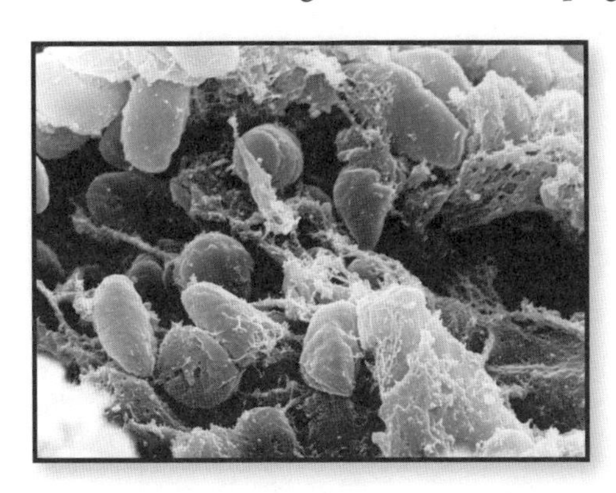

Scanning electron microscope depiciting a mass of Yersinia pestis bacteria in the foreground of a flea vector. Photo by Rocky Mountain Laboratories, NIAID, NIAI.

No precautions against the plague were prescribed in the Levitical Code. As the plague is not endemic in Palestine, the Jews probably incurred it by mixing with their neighbors. The Bible describes an outbreak of the disease among the Philistines in 1320 B.C. (1 Sam. 5:6, 9, 11). Also one must remember that in Deut. 28:60 the Jews were cursed for their disobedience with all the diseases of Egypt. Some have supposed emerods to be hemorrhoids, by comparison with the phrase Ps. 78:66, but this is doubtful. The same word emerods occurs in Deut. 28:27 and it states they cannot be healed. They lacked the antibiotics like streptomycin and tetracycline, which are effective against the plague bacillus. In 1 Sam. 5:6, 9, and 12 emerods are mentioned again where tumors or ulcers occurred in the anal region.

Plague outbreaks are most common in rural areas and in urban conditions characterized by overcrowding, poor sanitation and a high rat population. Outbreaks can occur at any time of year. Other animals that may be infected and pose a transmission risk to humans include wild rabbits and domestic cats that have contacted wild rodents. The disease usually spreads through fleabites, but one can also contract plague after being exposed to an infected, coughing animal or through a break in the skin after handling an animal with plague.

Currently the major threat is not through natural transmission of plague, but through intentional transmission, by terrorists. This bacterium was used in biological warfare in Europe in 1346, when the invading Mongol army catapulted plague-riddled cadavers into their enemy's city.

Consumption. Consumption is mentioned in Lev. 26:16. It is noted by wasting fever, characterized by weakness and anemia, often of long duration perhaps Mediterranean or Malta fever. Malta fever has already been discussed in detail where it is more certain that Malta fever is the disease being described. Also consumption is also known as tuberculosis, and this disease will be discussed later in this chapter in more detail.

Dysentery. This disease is characterized by frequent watery stools, often with blood and mucus, and characterized clinically by pain, tenesmus, fever, and dehydration. It may result from infection with *Entamoeba histolytica,* species of *Shigella* or viruses.

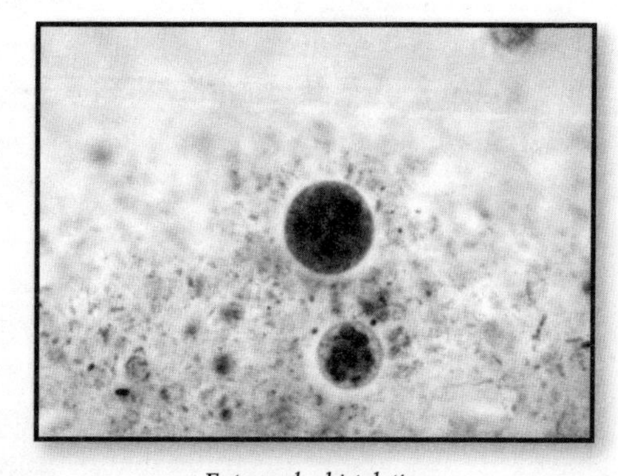

Entamoeba histolytica

In Acts 28:8 (bloody flux) is described which may be dysentery, which is prevalent in Malta. The mention of hemorrhage in this case shows that it was ulcerative, which is very dangerous. This dysentery is now called bacillary dysentery or shigellosis, which is caused, by *Shigella dysenteriae, S. flexneri, S. boydii, and S. sonnei.* The latter causes eighty percent of the cases in the United States. *Shigella* are facultative anaerobic gram-negative rods. Facultative anaerobic means the organism can grow aerobically in the presence of oxygen or anaerobically in its absence.

Shigella spreads rapidly in overcrowded conditions with poor sanitation. Humans and other primates such as apes and monkeys are the only reservoir of infection. These pathogens are spread by contaminated food, fingers, flies, feces, and fomites. An infective dose of just ten bacteria can cause disease. Thus, even slight lapses in hygiene can allow easy spread of the disease. The bacteria are not much affected by stomach acidity. They proliferate to immense numbers in the small intestine, but cause damage to the large intestine. The pathogens invade host cells and induce the cells to produce special filaments for invading adjacent host cells. After an incubation period of one to four days, abdominal cramps, fever, and profuse diarrhea with blood and mucus appear. The severity of signs and symptoms range from least serious, which could be so-called traveler's diarrhea caused by *S. sonnei,* to severe dysentery caused by *S. dysenteriae.* Along with fever-eliciting endotoxin this species produces Shiga toxin, which acts as a neurotoxin and inhibits protein synthesis, thereby killing cells. Infected people may have as many as twenty bowel movements in one day. This may have been what the father of Publius was infected within Malta at the time of Paul's shipwreck. This is one of those cases where symptoms could point to brucellosis, which has already been discussed.

Another dysentery is described in 2 Chron. 21:19 where Jehoram is smote by God in the bowels with an incurable disease. This condition appears to be amebiasis caused by *Entamoeba histolytica,* a major pathogenic amoeba. Amoeba are microscopic protozoans that are eucaryotic. Thus protozoans are in the Domain Eukarya. Amebiasis can appear as a severe acute disease called amoebic dysentery or as chronic amebiasis, which can suddenly revert to the acute stage. Human become infected with the parasite by ingesting cysts in food or water contaminated with fecal material. Flies and cockroaches can also be mechanical vectors. Cysts are a specialized cell enclosed in a wall. They are dormant, resistant structures, which by exact stimuli can excyst after ingestion by the host and form the vegetative form called the trophozoite. Trophozoites reproduce asexually within the colon. They may invade the intestinal mucosa and cause significant ulceration. Sometimes their proteolytic enzymes digest deep into, or even through, the bowel wall. Patients with amebiasis have abdominal tenderness, thirty or more bowel movement per day, and dehydration from excessive fluid loss. If the parasites invade liver and lung tissue, they can cause abscesses. Certainly this type of dysentery could last over two years.

Erysipelas. A fever or extreme burning (Deut. 28:22) could be some unspecified kind of irritating disease such as erysipelas (St. Anthony's fire). Erysipelas is a skin infection typically caused by group A beta-hemolytic streptococci, although other Lancefield groups are occasionally causative agents. Grouping is a serological classification scheme developed by Rebecca Lancefield in 1933 for differentiating beta-hemolytic streptococci. Group A streptococci are biochemically classified as *Streptococcus pyogenes. S. pyogenes* are 0.5 to 1.0 micrometer cocci that form short chains in clinical specimens. Growth is optimal on enriched blood agar media where they form large zones of beta-hemolysis or complete clearing of the red blood cells.

Erysipelas is an acute superficial cellulitis of the skin with prominent lymphatic involvement. Erysipelas occurs most commonly in

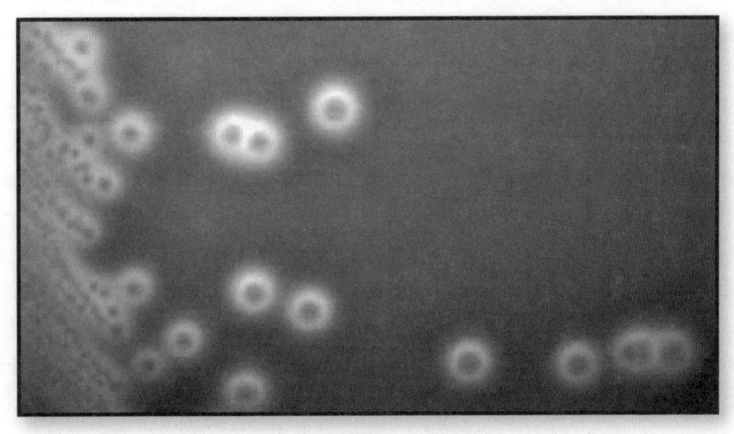

"Colonies of Streptococcus pyogenes on Blood agar"

young children or older adults, involves the face and less frequently the trunks or extremities, and usually is preceded by either respiratory or skin infections with *S. pyogenes.* The cutaneous manifestations are accompanied by chills, fever, and systemic toxicity. This form of the disease can be fatal when associated with bacteremia. Bacteremia is when the bacterium enters the bloodstream. The mortality rate is less than one percent in treated cases but would be much higher in bible times since no antibiotics were available.

Fever – Malaria. Fever is a bodily temperature above the normal of 98.6 (37° C), a disease in which there is an elevation of the body temperature above the normal. Several forms of febrile disease were among the commonest of maladies in the period covered by the Bible history. In Lev. 26:16 fever is mentioned and can be translated by the Greek word for jaundice. It may be malarial fever (ague), which occurs in certain parts of Palestine and is occasionally accompanied by jaundice. This may be the disease alluded to in Matt. 8:14 and Luke. 4:38, where Peter's wife mother had the disease. Perhaps Jesus employed suggestive mechanisms in therapy to lower the temperature (Luke. 4:39). Perhaps the illness that afflicted the nobleman's son (John: 49-53) was a febrile condition, possibly malaria.

Malaria is a severe parasitic disease caused by several species of the protozoa *Plasmodium* and transmitted by mosquitoes. The protozoa are

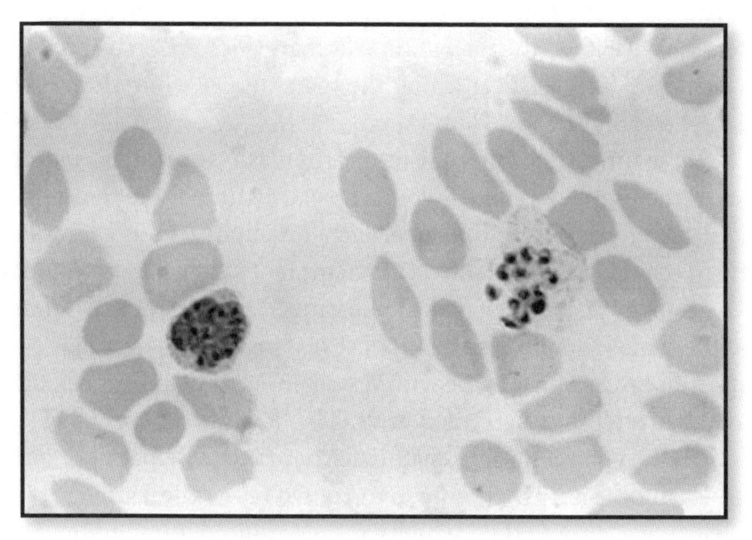

"Plasmodium in human blood smear"

animal-like, mostly unicellular microbes. In any given year worldwide it causes 300-500 million cases and claims the life of 1.5 to 3 million a year, many of them children. Malaria is an ancient human scourge. Written records on Egyptian papyrus from 1550 B.C. describe a disease with high intermittent fever that must have been malaria. In some low-lying ancient cities, nearly entire populations succumbed to the disease. Although humans were unaware of the origin of malaria and its transmission, protective measures against the disease were used for hundreds of years. The inhabitants of swampy regions in Egypt were recorded as sleeping in tower-like structures out of the reach of mosquitoes, whereas others slept under nets as early as 450 B.C.

The female *Anopheles* mosquito bites an infected person, becomes infected, then transmits the parasite to the next individual it bites. In the body, the parasites destroy the red blood cells, causing a wave of intense cold followed by intense fever, the so-called malaria attack. Extensive anemia develops, and the hemoglobin from ruptured red blood cells make the urine dark (giving rise to the name "blackwater fever"). Often vomiting and severe headaches occur.

Malaria is now diagnosed by clinical symptoms and microscopic examination of the blood.

Gonorrhea. In Lev. 15 the Bible speaks of men and women being unclean by their issues (bodily discharge). The discharge might be diarrhea or urethral discharges caused by various types of microbial infections. Any bed the man with a discharge lies on or sits on will be unclean (Lev. 15:4). Everything the man sits on while riding is unclean (Lev.15:9).

In Bible times, like current times, people had sexual temptation. In Prov. 7 the first five verses, Solomon pleads with his son to listen and guard his words carefully concerning the adulteress. If one follows Lev. 18 that gives us the laws of sexual morality, then STD's would not be a problem in any society. In Ephesians 5:3 God says that, "Among you there must not be even a hint of sexual immorality. Sexually transmitted diseases (STD) come with sexual impurity. It appears that King David had a STD as described in Ps. 38:1-8. Some of these STDs are gonorrhea, syphilis, herpes, chlamydia, hepatitis C. hepatitis B, or Human Papilloma Virus. Of course today we know the Human Immunodeficiency Virus (HIV), which causes Acquired Immunodeficiency

Syndrome (AIDS) is an STD. So to narrow the topic only gonorrhea and syphilis will be discussed.

Gonorrhea is a STD that affects the mucosa of the reproductive tract of men and women. It is caused by a small gram-negative diplococcus commonly known as the gonococcus, and has a worldwide distribution. The bacteria are phagocytized by epithelial cells lining the reproductive tract and grow inside the cells. Eventually the organisms rupture the cells, thereby penetrating the epithelia barrier. Phagocytes, such as neutrophils, also may contain gonococci inside vecsicles. Eventually the organisms rupture the cells, thereby penetrating the epithelial barrier. The inflammatory response evoked leads to production of a vaginal or penile discharge, the primary symptoms of gonorrhea.

In males the incubation period is two to eight days. The onset consists of a urethral discharge and frequent painful urination that is accompanied by a burning sensation. In females the disease is sometimes asymptomatic. However, some symptoms may begin seven to twenty-one days after infection. These are generally mild and some vaginal discharge may occur. The gonococci also can infect the uterine tubes and surrounding tissue, leading to pelvic inflammatory disease (PID). Gonococcal PID is a major cause of sterility and ectopic pregnancies. In both sexes disseminated gonococcal infection with bacteremia may occur. This can lead to involvement of the joints (gonorrheal arthritis). Gonococcal pharyngitis may develop in the pharynx. Gonococci may cause eye infections most often in newborns as they pass through an infected birth canal. This disease termed ophthalmia neonatorum or conjunctivitis of the newborn, which was once a leading cause of blindness. In John 9:1, a person was born blind and this affliction could have been from ophthalmia. Now to preclude blindness in the U.S., most states have laws requiring that the eyes of newborns be treated with silver nitrate or antibiotics. Antibiotics are preferable since strains of *Chlymdia trachomatis* can also cause this disease.

Syphilis. This disease and other STDs are exemplified in the truth of the Bible warning of the visitation of the iniquity of the fathers upon the children, and upon the children's children, unto the third and to the fourth generation (Ex. 34:7). Syphilis is a chronic systemic infection caused by *Treponema pallidum. T. pallidum* is a thin, delicate bacterium with six to fourteen spirals and tapered ends measuring six to fifteen micrometers in total length and 0.2 micrometers in width. The organism is difficult to study because it does not grow on laboratory media.

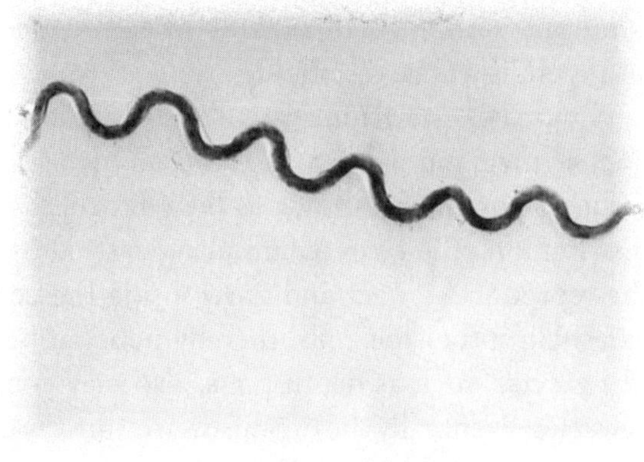

Treponema pallidum

Syphilis is transmitted by sexual contact of all kinds, via syphilitic infections of the genitals or other body parts. The incubation period averages three weeks, but ranges from two to six weeks. The disease progresses through several recognized stages. The early stage of syphilis (primary syphilis) is characterized by a lesion, called a chancre, that appears at the site of infection. Not all people develop a chancre, and some people who develop one do not notice it because it is hidden in the vagina, mouth, or anus. The chancre is painless, and an exudate of serum forms in the center. This fluid is highly infectious, and examination with a darkfield microscope shows many spirochetes. The chancre disappears spontaneously after a couple of weeks. The bacteria rapidly penetrate mucosal membranes and enter the blood stream and lyhmphatic system, which distributes them widely in the body.

Several weeks after the primary stage, the disease enters the secondary stage, characterized mainly by skin rashes of varying appearance. Painful, whitish mucous patches swarming with spirochetes appear in the tongue, cheek, and gums. Inside the body the treponemes invade the heart, musculoskeletal system, and central nervous system. The disease then enters another stage called the latent period in which the bacteria are difficult to find. After two to four years of latency, the disease is not normally infectious, except for transmission from mother to fetus. In less than half of untreated cases, the disease reappears in a tertiary stage. Symptoms of tertiary syphilis develops over a period of years to decades and include heart damage, neurological symptoms, disabling fatigue, and disfiguring skin lesions. The hallmark of tertiary syphilis is the gummas, a soft, gummy granular lesion that weakens the blood vessels, causing them to burst. In the central nervous system, gummas lead to paralysis and insanity.

In Zech. 11:17, a curse was invoked upon the idol shepherd, one result of which would be his arm would shrivel up completely. This is possibly a reference to locomotor ataxia, a disease of the spinal cord resulting in chronic progressive sclerosis of the posterior spinal nerve roots and accompanying structures. It is generally the result of syphilis, and the disease normally commences in middle age. The locomotor ataxia condition results in marked diminution of muscular coordinating power followed by paralysis. Atrophy of the optic nerves (Zech. 11:11b) may occur at an early stage of the disease. Locomotor ataxia may pursue a chronic course, or it may have a rapid fatal termination.

Inflammation – typhoid fever. Inflammation is a pathological process consisting of a complex of cytologic and histologic reactions that occur in the affected blood vessels and adjacent tissue of man and other animals; it is in response to an injury or abnormal stimulation caused by a physical, chemical, or biologic agent or combination of such agents. Inflammation is characterized by pain, swelling, redness and heat. Thus inflammation could be a bacterial disease such as typhoid fever, which is common in Palestine (Deut. 28:22; Lev. 13:28). Typhoid is common in some crowded towns and villages. In those times little was done to protect the wells from contamination. Perhaps the plague that destroyed Sennacherib's army in 2 Kings 19:7 was typhoid fever.

Salmonella typhi causes a life-threatening illness called typhoid fever. This bacterium is a straight rod, 0.7-1.5 by 2.0-5.0 micrometer and possesses peritrichous flagella for motility. After a ten to fourteen day incubation period following ingestion of the bacilli, the patient has a gradually increasing remittent fever as high as 103° to 104° F (39° to 441° C) with nonspecific complaints of stomach pain, headache or loss of appetite. Some patients will have a rash of flat, rose-colored spots. Chronic carriers for more than one year after symptomatic disease will develop in one to five percent of patients, with the gall bladder serving as the reservoir in most patients.

In the United States, about four hundred cases occur each year and seventy percent of these are acquired while travelling internationally.

Typhoid fever is still common in the developing world, where it affects about 12.5 million persons each year.

S. typhi lives only in humans. One can get typhoid fever by eating food or drinking beverages that have been handled by a human who is shedding *S. typhi.* It is also obtained if sewage contaminated with *S. typhi* gets into the drinking water or in food washed with contaminated water.

Therefore, typhoid fever could have been common in Bible times where hand washing might have been less frequent and their drinking water possible contaminated with sewage.

Itch. Itch or pruritus has many causes. There are many skin diseases that itch, but the one generally referred to by the term " the itch" is scabies. This is probably the parasitic disease referred to in Deut. 28:27 which is due to a small mite which burrows under the skin, and, if neglected, sometimes spreads all over the body. This disease is easily transmissible. It is usually found in people with unclean habits who seldom bathe and who wear unclean clothes. Scabies is spread from person to person by intimate personal contact and is facilitated by crowding, uncleanliness, and sexual promiscuity.

Probably the first documented connection between a microscopic organism and a disease in humans was scabies, which was described by an Italian physician in 1687. Scabies, or sarcoptic mange, is caused by the itch mite *Sarcoptes scabiei. S. scabiei* burrow under the skin to lay its eggs. By the time intense itching appears, the lesions usually are quite widespread. Scratching lesions and causing them to bleed provides an opportunity for secondary bacterial infections.

The human itch mite, which infects some 300 million persons each year, is one of the most common causes of itching dermatoses throughout the world. Burrows appear as dark wavy lines in the epidermis measuring 3-15 mm and ending in a small pearly bleb, which contain the female mite. In the majority of patients they occur on the volar wrists, between the fingers, on the elbows and on the penis. Small papules and vesicles, often accompanied by eczematous plaques, pustules, or nodules, are symmetrically distributed in these sites and in skin folds under the breasts and around the navel, axillae, belt line, buttocks, upper thighs, and scrotum.

In Lev. 21:20 it states that the itch was a disqualification for the priesthood. Another group of skin diseases that itch are those caused by fungi (molds). The basic characteristics of eucaryotic molds are structural units, which are tube-like projections known as hyphae. As the hyphae grow, they become intertwined to form a loose network, the mycelia, which penetrate the substrate from which the mold obtains the necessary nutrients for growth. Vegetative hyphae comprise the body of the fungus, in contrast to specialized reproductive hyphae. The nutrient absorbing and water-exchanging portion of the fungus is called a vegetative mycelium. The portion extending above the substrate surface is known as aerial mycelium. Aerial mycelia often give rise to fruiting bodies from which asexual spores are borne.

Probably the microbe mentioned in Lev. 21:20 is ringworm. Despite its name, ringworm has nothing to do with worms but fungi. Fungi that colonize the hair, nails, and the outer layer of the epidermis are called dermatophytes; they grow on the keratin present in this location. The genera *Trichophyton, Microsporum* and *Epidermophyton* are the principal etiologic agents of the dermatomycosis. Such cutaneous mycosis are perhaps the most common fungal infection of humans and are usually referred to as tinea (Latin for "worm" or "ringworm") The gross appearance of the lesion is that of an outer ring of an active, progressing infection with central healing within the ring. These infections may be characterized by another Latin noun to designate the area of the body involved. For example, tinea corporis (body), tinea cruris (groin), tinea capitis (scalp and hair), tinea barbae (beard), and tinea unguium (nail) are used to designate the type of infection produced. There are more than forty species of dermatophytes, so it is difficult to specifically know which dermatophyte the Bible is speaking about. Dermatophytic infections are most common in tropical climates and under crowded living conditions. Some dermatophytes live on animals which the Bible speaks of in Lev. 22:22. In Lev. 13:29 it speaks of tinea barbae. In Lev. 13:42 the leprosy mentioned might be tinea favasa; a chronic fungus infection of the scalp and nails caused by *Trichophyton schoenleinii. T. schoenleinii* causes a severe type of infection called favus. It is characterized by the formation of yellowish, cup-shaped crusts, considerable scarring of the scalp; and sometimes permanent alopecia (baldness).

Leprosy (Hansen's Disease). Leprosy is an ancient disease described in the Bible and feared throughout history as contagious and disfiguring. The Hebrew word leprosy was used for various diseases affecting the skin—not necessarily leprosy. Thus in this section of the book only verses of the Bible will be cited where the patient appears to have bacterial leprosy.

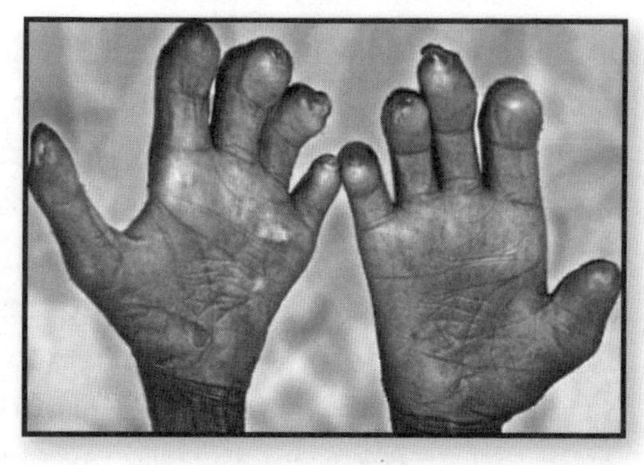

Treponema pallidum

In 2 Kings 5:1 it states that Naaman, commander of the Syrian army had leprosy. In 2 Kings 5:9-14, God working through Elisha healed him of this dreadful disease. In 2 Kings 7:3-5 four men suffering from leprosy went to surrender to the Syrian army but found their camp abandoned. In Lev. 13:1-17 the laws were given to Moses and Aaron discerning leprosy.

Matt. 8:1-17 records a number of miracles demonstrating Jesus' concern for those who suffer physically. It is interesting that the first miracle by Jesus was the healing of a leper. To be a leper was considered the most terrible fate that could befall anyone. It was better to be dead than to be a victim of leprosy in ancient time. A leper was subject to Old Testament laws, which made him an outcast, both socially and spiritually. Cut off from all normal contact with society, he was forced to live apart from his family. He had to wear torn clothing, which identified him as a living corpse. Wherever he went, he had to cry, "Unclean! Unclean!" so that people would know to stay away from him.

From the touch of Jesus recorded in Matt 8:3, the leper was healed. This also recorded in Luke 5:13. Jesus then indicated his approval of the rule God had given through Moses concerning the healed leper. He was to go to the priest, show evidence of his cure, and then present two birds, one to be killed and the other to be released as a par of the disinfection ceremony (Lev. 14:4-7).

It is interesting that in Luke 17:11-19 Jesus heals ten men who suffered from leprosy, but only one returned to thank Him. May it be our prayer today to always be thankful for the way God ministers to us.

The causal agent of leprosy, *Mycobacterium leprae,* has the potential for causing disfigurement. It is closely related to *M. tuberculosis,* the microbe that causes TB; both have a complex cell wall that is stained by the acid-fast technique. Infected individuals who show maximum resistance to the bacilli demonstrate a disease affecting the superficial nerve endings and related skin areas. In persons with minimal resistance, the bacilli may be disseminated throughout the body.

M. leprae is an intracellular parasite that has never been grown in bacteriological media or cell culture, but can be cultured in mouse footpads or armadillos. Like most other mycobacteria, *M. leprae* exhibits slow growth and has a generation time of twelve days when cultivated in armadillo or mouse footpads.

Worldwide, one to two million persons are permanently disabled as a result of leprosy. However, persons receiving antibiotic treatment or having completed treatment are considered free of active infection.

Transmission of the leprosy bacillus is believed to occur through inhalation of bacilli onto nasal mucosa, intact skin, or penetrating wounds such as by thorns. Inhalation of aerosols generated by infected patients is believed to be the most common method of transmission. Leprosy is a natural infection of armadillos, but whether these animals are a source of infection to humans is not certain since most leprosy cases occur where armadillos are absent.

Most people in areas endemic for leprosy become infected without any overt symptoms being manifested. Disease symptoms appear from two to four years after infection. In one form of the disease, called indeterminate leprosy, a few hypopigmented areas of the skin plus dermatitis may be the first clinical manifestation of disease. Most (75%) of these individuals will recover spontaneously but the remainder will progress to one of the established forms of the disease, called lepromatous leprosy or tuberculoid leprosy.

Lepromatous leprosy is the most disfiguring form of the disease because of the host's lack of immunity to the bacillus. Skin lesions range from diffuse to nodular. The skin lesions appear in the cooler parts of the body (nasal mucosa, anterior one-third of the eye, and peripheral nerve trunks at specific sites: elbow, wrist, hands and ankle). The patient in lepromatous leprosy has few symptoms other than those caused by the nodular masses. Facial features become thickened, taking on the classical leonine (lion-like) appearance of leprosy. The nose may collapse because of extensive tissue destruction. Fingers or toes may be lost. In advanced cases there is sensory loss due to involvement of nerve fibers.

Patients with tuberculoid leprosy mount a vigorous cell-mediated immune response that holds the infection in check. There is a localized infection in which there are a few well-circumscribed skin lesions. The lesions appear flat, are blanched, and contain few bacilli.

In most leprosy patients there is some degree of irreversible peripheral nerve damage.

In advanced cases, widespread destruction of nerve trunks can result in loss of feeling and permanent paralysis involving face, hands, and feet. The lack of pain sensation allows the patient to suffer self-damage and self-deformation through continued use of an extremity. Since leprosy causes a variety of signs and symptoms certainly many of the verses of the Bible dealing with leprosy could have been caused by the Hansen's bacillus.

Parasitic Worms. Luke describes the last days of Herod Agrippa in Acts 12:21-23 where Agrippa had probably suffered intestinal obstruction, perhaps because of the presence and activity of parasitic worms. He may have died from perforation of the bowels, with resultant peritonitis. Parasitic worms sometimes leave the body after the individual has died, and this may have been the reason for their mention in Luke's account. On the other hand, they could have been voided while Agrippa was still alive smitten by a terrible disease by God; Agrippa dies in great agony.

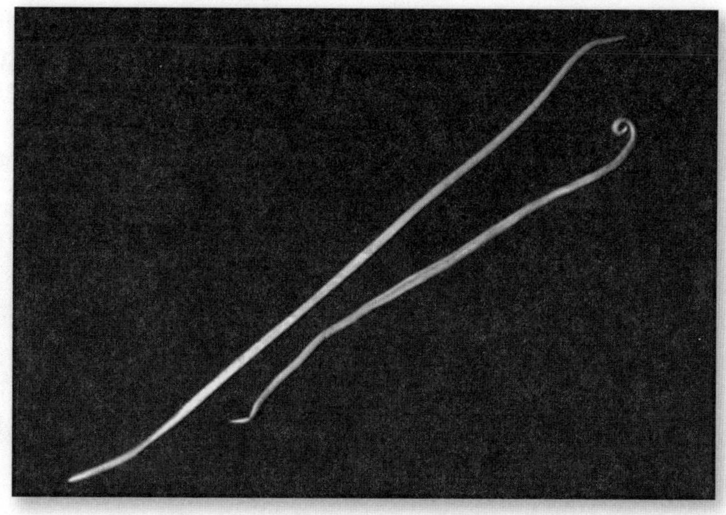

Ascaris lumbrocoides

The helminth that might have caused Agrippa's death is *Ascaris lumbrocoides. A. lumbricoides* is one of the most cosmopolitan worms worldwide. Ingested eggs releases a larval worm that penetrates the duodenal wall, enters the bloodstream, is carried to the liver and the heart, and then enters the pulmonary circulation. The larvae breaks free in the alveoli of the lungs, where they grow and molt. In about three weeks the larvae pass from the respiratory system to be coughed up, swallowed, and returned to the small intestine.

As the male and female worms mature in the small intestine (primarily jejunum), fertilization of the females by the males initiates egg production, which may amount to 200,000 eggs per day for as long as a year. The eggs are round or ovoid with a thick shell, are brown or yellow brown and are corticated. Female worms can also produce unfertilized eggs in the absence of males. Eggs are found in the feces sixty to seventy days after the initial infection. Fertilized eggs become infectious after approximately two weeks in the soil.

Lumbricoides is prevalent in areas of poor sanitation and where human feces are used as fertilizer. Because both food and water are contaminated with *Ascaris* eggs, the parasite more than any other affects the world's population. Although no animal reservoir is known for *A. lumbricoides, A. suim* can infect humans.

Infections that are caused by the ingestion of only a few eggs may produce no symptoms; however, even a single adult ascaris may be dangerous because it can migrate into the bile duct and liver and create tissue damage. Furthermore, since the worm has a tough flexible body, it can occasionally perforate the intestine, creating peritonitis with secondary bacterial infection.

Following infection with many larvae, migration of worms to the lungs can produce pneumonitis resembling an asthmatic attack. Also, a tangled bolus of mature worms in the intestines can result in obstruction, perforation, and occlusion of the appendix. Patients with many larvae may also experience abdominal tenderness, fever, distention, and vomiting. Occasionally adult worms pass with the feces, which can be quite dramatic because of their large size (20-30 cm long).

There are several poisonous serpents in the desert of the Exodus narrative, whose bites are often fatal. It has been suggested that the fiery serpents of Num. 21:6 were really the parasitic worms called guinea worms, which are not uncommon in the desert region. Guinea worms have been traced to calcified worms in the stomach of Egyptian mummies during the first millennium. *Dracunculiasis* (guinea worm infection), caused by *Dracunculus medinensis* is a tissue-invading nematode. The worms have a very simple life cycle, depending on fresh water and a microcrustacean (copepod) of the genus *Cyclops.* When Cyclops harboring larval *D. medinensis* are ingested in drinking water by humans and other mammals, the infection is initiated with liberation of the larva in the stomach. Larvae penetrate the stomach or intestinal wall, mate, and mature. The adult male probably dies; the female *Dracunculus* develop over a year and migrate to subcutaneous tissues, usually in the lower extremity. As the thin female *Dracunculus,* ranging in length from 300cm to 1 m, approaches the skin, a blister forms over days, breaks down and forms an ulcer. When the blister opens, large numbers of motile rhabditiform larvae can be released in the stagnant water; ingestion by the *Cyclops* completes the life cycle.

Symptoms of infection usually do not appear until the gravid female creates the vesicle and the ulcer in the skin for liberation of larval worms. At the site of the ulcer there is erythema and pain, as well as an allergic reaction to the worm. There is also the possibility of abscess formation and secondary bacterial infection, leading to further tissue destruction and inflammatory reaction, with intense pain and sloughing of skin.

If the worm is broken in attempts to remove it there may be toxic reactions, and if the worm dies and calcifies there may be nodule formation and some allergic reaction. Occasionally, the adult worm does not emerge but becomes encapsulated and calcified.

Polio. The man with the withered hand (Matt. 12:10-13; Mark 3:1-5; Luke 6:6-10) may have suffered from poliomyelitis as a youth. The absence of detailed description only leads one to guess what caused the withered hand.

The impotent man at Bethesda (John 5:2-9) appears to be the victim of poliomyelitis as a child. After Jesus healed him he is now able to walk.

Poliomyelitis is a very ancient disease; its effects are clearly depicted in an Egyptian hieroglyph in 2300 B.C. Polio is caused by a virus. Unlike bacteria, molds, yeast, and algae, viruses are not cells. Thus they are not procaryotic or eucaryotic. They have unique structural and functional properties and cannot reproduce unless they are inside a cell. In fact, by themselves, viruses cannot generate energy or make proteins. The common thread between viruses and cells is nucleic acid. Viruses have only one type of nucleic acid, either RNA or DNA, but not both. Surrounding the nucleic acid is a protective protein coat, or capsid. Together, the nucleic acid and protein capsid are called the nucleocapsid. The whole acellular viral particle is called a virion. All virions have a nucleocapsid and for many, including polioviruses, the nucleocapsid constitutes the entire virion.

Polioviruses are small RNA viruses with a diameter of approximately 25 nm, they are some of the smallest viruses. There are three antigenic distinct polioviruses designated type 1, 2, and 3.

Polio is primarily a gastrointestinal disease, but polioviruses can spread from the intestine to the central nervous system. Polio varies in severity from an asymptomatic condition, to a mild febrile illness, to a paralytic disease that may result in death. When motor neurons (nerve cells) in the gray matter of the spinal cord are killed by polioviruses, the muscles they control cannot move. Muscles that do not function will atrophy, giving rise to one of the distinctive signs of paralytic polio in polio survivors, a withered leg that has strikingly smaller diameter than its mate which is clearly observed in the Egyptian hieroglyph.

The gastronintestinal form of polio, with which the disease begins, is characterized by fever, malaise, headache, nausea, and vomiting. Patients appear to be recovering but then sometimes the disease progresses to include muscle pain, stiffness of the neck and back, and flaccid (limp) paralysis. If muscles involved in breathing or swallowing are affected, then polio becomes life-threatening.

Polioviruses are in the enterovirus subgroup of the picornaviruses. All the enteroviruses infect cells of the gastrointestinal tract. Consequently, they are shed and transmitted mainly in fecal matter and oral secretions. Polioviruses are more stable than most other viruses and can remain infectious for relatively long periods in water and food. The primary mode of transmission is ingestion of water contaminated with feces containing the virus.

Scab. Scab is an eschar; a crust formed by the drying of the pus on the surface of an ulcer or linear break in the skin surface, usually covered with blood or serous crusts. This term could be referring to scabies in Lev. 21:20 or be referring to tinea as in Isa. 3:17. In Deut. 28:27 the scab could be referring to eczema, tinea, syphilis, impetigo or leprosy.

Scall or scurf. This term means a pustular scaly eruption of the skin, beard or scalp. In Lev. 13:30 the scall of the skin or scalp is referring to tinea. Scurvy mentioned in Lev. 21:20 probably means scaly or scabby and does not refer to the disease entity of scurvy with hemorrhages of skin and mucous membranes due to a monotonous diet of salt meals and the lack of fresh vegetables, fruits and other sources of vitamin C in the diet.

Smallpox. The descriptions given in Ps. 39:11, Zech. 14:12, Ps. 38:5, Lev. 26:29, and Ezk. 24:23; 33:10 are largely figurative, but the imagery may be taken from an attack of smallpox, with its disfiguring and repulsive effects.

If fact the plague that God did against Pharaoh, the 6th plague of festering boils on people and animals could have been smallpox (Ex. 9:8-12). It appears God used biological warfare on Pharaoh's kingdom and the Israelites were immune.

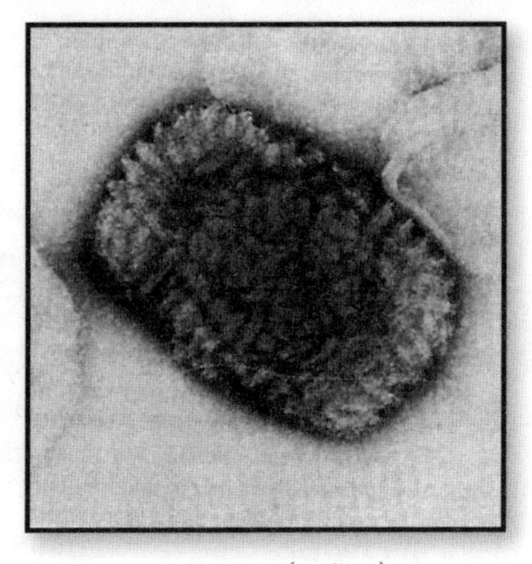

Variola Minor (smallpox)

Smallpox is an acute contagious disease caused by variola major, a member of the Poxviridae family. Variola is a large (200 to 400 nm) DNA virus with complex structure and appears brick-shaped. The incubation period is seven to nineteen days in a typical case of ordinary type smallpox. The virus causes fever, a vesicular and pustular eruption with a high mortality rate. A milder form (variola minor) causes a milder course with a mortality rate of less than one percent.

The disease first appeared sometime after 10,000 B.C. in a small agriculture settlement in Asia or Africa. The mummy of Rames V, who died in Egypt in 1160 B.C., had smallpox scars on the face, neck, shoulders, and arms. Smallpox ravaged villages in India and China for centuries, and in epidemic occurred in Syria in 302 A.D. Smallpox was the first virus to be studied in detail and was the first virus for which a vaccine was developed in 1796, by Edward Jenner of England. His residence is still available for touring. He showed that inoculation with cowpox virus protected against smallpox. Vaccinating with this vaccinia virus gave protection against smallpox for five years and some protection for twenty years. Periodic revaccination is necessary for optimal protection. In 1967, the World Health Organization adopted a program aimed at global eradication of smallpox. Through this eradication program the last natural case of the disease occurred in 1977. This was the first infectious disease to be eliminated in this fashion. Because humans are the only hosts, the spread of smallpox was prevented until no new cases developed.

Spinal Meningitis. Palsy or paralysis mentioned in Matt. 8:6, 9:2, could have been spinal meningitis. There are many types of palsy such as facial or that which occurred during birth. Certainly microbes can cause paralysis.

Meningitis is an infection of the tissues (meninges) and sometimes the fluid (cerebral spinal fluid) that surrounds the brain and spinal cord. Meningitis results in swelling of the brain tissue and in some cases the spinal tissue (spinal meningitis). When brain tissue swells, less blood and oxygen reach brain cells. The infection occurs most often in infants, young adults between the ages of 15-24, older adults, and people who have a long standing health condition such as a weakened immune system. Meningitis can range from mild to life-threatening. The severity usually depends on the microbe causing the infection and a person's age and overall health.

Meningitis can be caused by different types of pathogens, including viruses, bacteria, fungi, and protozoa. Viral meningitis, mostly caused by echoviruses, is probably more common than bacterial meningitis but tends to be a mild disease. The most common cause of bacterial meningitis are *Neisseria meningitidis* (a gram-negative bacteria that causes epidemic meningitis, *Streptococcus pneumoniae* (a gram-positive diplococcus that is also a common cause of bacterial pneumonia), and *Haemophilus influenzae* type b (a gram-negative bacterium which was the most common cause of meningitis in children until the HIB vaccine was approved for use in 1990). These three bacterial species cause more that 70% of meningitis cases.

What do these three diverse types of diverse bacteria have in common? All three posses a polysaccharide capsule that protects them from phagocytosis as they multiply in the bloodstream, from which they might enter the cerebrospinal fluid. Death from bacterial meningitis often occurs quickly, probably from shock and inflammation. Nearly fifty other species of bacteria have been reported to be opportunistic pathogens that occasionally cause meningitis.

Toxic Algal Blooms. The Bible states "all the waters that were in the river turned to blood, and the fish that was in the river died (Ex. 7:20-21)" causing a terrible stench. This is the first plague Moses visited on the Egyptians. All the water in Egypt—right from water already in buckets and jars, to ponds, canals, streams and even the Nile River—turned to blood. The Red Sea probably is named after these toxic algal blooms. The poisonous and destructive red tides that occur frequently in costal areas often are associated with population explosions, or "blooms" of dinoflagellates. The pigments in the dinoflagellate cells are responsible for the red color of the water. Under these bloom conditions, the dinoflagellates produce a powerful neurotoxin called saxitoxin. This toxin is excreted into the water. This could have occurred at the time of the lowest Nile just before the rise occurs. Then the water becomes defiled and very red, so polluted that the fish die. This time period would be some period in the month of May. The severity of the plague, constituted the "wonder" in the first plague.

When humans ingest shellfish such as oysters and clams which themselves ingested toxic dinoflagellates, manifestations include numbness of the mouth, lips, face, and extremities, and diarrhea and occasionally paralysis and respiratory arrest. Even animals as large as dolphins have been killed in large numbers by this toxin. Inhaling air that contains small quantities of the toxin can irritate respiratory membranes, so sensitive individuals should void the sea and its products during red tides.

Tuberculosis. Nearly one-third of the world's population is infected with tuberculosis (TB), which kills almost three million people per year. The disease is also known as the great white plague or "consumption". The word consumption is employed in the bible as Schachepheth, which is also the Modern Hebrew word tuberculosis.

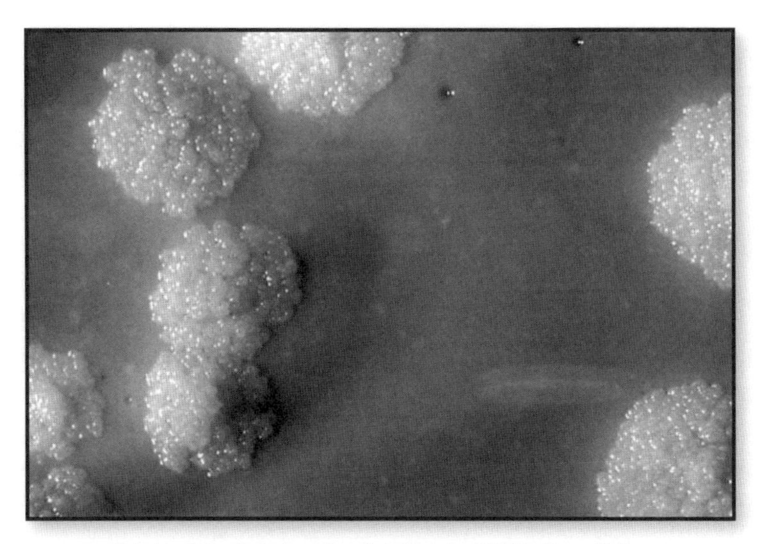

Colonies of Mycrobacterium tuberculosis. Photo by CDC/Dr. George Kubica.

TB is an infectious disease caused by the bacterium *Mycobacterium tuberculosis,* a slender rod and an obligate aerobe. The rods grow slowly sometimes forming filaments, and tend to grow in clumps. They have a 20-hour generation time or longer compared to many other prokaryotic organisms that have a 20 minute generation time; the generation time is the time it takes the microbe to divide.

These bacteria are relatively resistant to conventional simple staining or the Gram's stain. Cells stained with carbol fuschin dye cannot be decolorized with acid-alcohol and are therefore classified as acid-fast. The acid-fast stain binds strongly only to bacteria that have a waxy material in their cell walls particularly composed of mycolic acid. Microbiologists use the stain to identify all bacteria in the genus *Mycobacterium.* Most bacteria are non-acidfast and are not stained by carbol fuschin but are stained blue by the counterstain methylene blue.

Tuberculosis can involve any organ, although most infections are restricted to the lungs. Infection is initiated after inhalation of contaminated aerosol droplets. If the infection progresses, the host isolates the pathogen in a walled-off lesion called a tubercle (meaning lump or knob), a characteristic that gives the disease its name. The onset is generally insidious, with nonspecific complaints of malaise, weight loss, cough, and night sweats. Sputum production may be scant or bloody and purulent. If the disease is arrested at this point, the lesions slowly heal, becoming calcified. If the body's defenses fail at this stage, the tubercle breaks down and releases virulent bacilli into the airways of the lung and then the cardiovascular and lymphatic systems. The disseminated infection is called miliary tuberculosis (miliary is derived from the numerous millet seed-sized tubercles formed in the infected tissues). TB in HIV-infected patients is twice as likely to spread to extropulmonary sites, and rapidly progress to death.

Lev. 26:16 speaks of consumption or TB as does Deut. 28:22. It is interesting that the well-known theologian John Calvin died of TB on May 27, 1564.

Tularemia. Ancient health keys revealed in the Bible shows us the way to avoid many illnesses and diseases. In Lev. 11:6-8, which in part states and of the hares, of their flock shall ye not eat, and their carcasses shall ye not touch, they are unclean to you. This would help keep the Israelites from being infected with the bacterium *Francisella tularensis.* This bacterium causes the disease tularemia, which is caused by a small, gram-negative, falcultative anaerobic, pleomorphic, rod-shaped bacteria. The microbe was named for Tulare County California, which it was originally isolated in 1911 from ground squirrels that died from tularemia. Today we know that tularemia occurs worldwide among wild mammals. *F. tularensis* commonly infects rabbits, squirrels, muskrats, and deer. Ticks are also a significant reservoir; they can transmit the bacterium transovarially (through the eggs) from one generation to the next.

Tularemia, or rabbit fever, exhibits one of several forms, depending on whether the infection is acquired by inhalation, ingestion, bites, or most commonly, contact through minor skin breaks. The most common clinical syndrome is ulceroglandular tularemia. This results in a small sore appearing where the bacteria entered the body. About a week after infection the regional lymph nodes enlarge; many will contain pockets filled with pus. If the disease is not contained by the lymphatic system, the microbes can produce septicemia, pneumonia, and abscesses throughout the body. Ingesting infected, inadequately cooked meat can lead to a focus of infection in the mouth or throat. The highest mortality rates occur with pneumonic tularemia.

Humans most frequently acquire the infection through minor skin abrasions and by rubbing the eyes after handling small wild mammals.

As few as ten bacteria introduced into a wound will initiate an infection. We know that about ninety percent of the cases in the U.S. are contracted from rabbits, and that was probably the case in Bible times.

In Lev. 11:7-8 and Deut. 14:8 it also says not to eat the swine since it is unclean to you. The contempt felt for swine is noted in Prov. 11:22 and Matt. 7:6. Pigs are unclean and they're susceptible to a greater number of diseases than any other domestic animal. Respiratory and parasitic ailments of pigs are major problems, particularly with limited exercise and lack of sunlight. Thus it would be beyond the scope of this book to discuss all the diseases that the Jews could of obtained by eating the flesh of swine. Certainly some swine diseases are transmissible to humans. Among them are brucellosis, trichinosis, and cysticerosis. Even today in the U.S. three of the most common food-borne parasitic diseases of humans are with pork consumption, which are toxoplasmosis, taeniasis (tape worm) and trichinosis.

Certainly *F. tularensis* is one of many pathogens that could be used effectively in bioterrorism.

Additional Reading:

Cieslak, T. J., and E. M. Eitzen. *Clinical and Epidemiologic Principles of Anthrax*, Emer. Inf. Dis. 5:552-555, 1999.

http://www.bible.wiktel.com/ntb/ntb00830.htm

http://www.medicinenet.com/Boils/article.htm

http://health-pictures.com/eryeipelas.htm

http://www.cdc.gov/travel/malinfo.htm

http://hlunix,hl.state.ut.us/els/epidemiology/epifacts/typhoid.html

http://100777.com/node/501

http://www.remnantofgod.org/7plagues.htm

Biblical Methods Used to Control Microbial Growth

Controlling microbes has always been of paramount importance. Microbes have the ability to grow and multiply at extremely high rates. This is a problem when pathogenic microbes grow in our bodies; it is a hazard to good health and must be controlled. This is also a problem when unwanted organisms grow in our food and cause spoilage.

On a grand scale, the development of human civilization has depended in large measure on the ability to manage the microbial world. We have needed to control microbes so as to prevent the epidemics that regularly spread through our population.

It is the purpose of this chapter to present mechanisms of control of microbial populations that the ancient Hebrews had at their disposal. Although microorganisms were not discovered until the 17th century and it was the 18th century before experimentally microbes were proven to cause disease, the Hebrews were treating diseases in the Old Testament times.

Sodium Chloride. Sodium chloride or common salt is the chemical compound NaCl. The first written reference to salt is found in the Bible in the Book of Job. There are thirty-one other references to salt in the Bible. The most familiar probably is the story of Lot's wife, who was turned into a pillar of salt when she disobeyed the angel and looked back at the wicked city of Sodom.

Salt has greatly influenced the political and economic history of the world. Roman soldiers were paid "salt money", salarium argentium, from which we take our English word, "salary". The Israelites were quite familiar with salt as they used it for seasoning and preservation of food (Job 6:6, Matt. 5:13, Mk 9:50 and Col. 4:6). They certainly used it for preservation of foods such as fish (Num. 11:5).

All meat-offerings were seasoned with salt. They also used it as an antiseptic when the midwife delivering an infant, severed its umbilical cord, washed the child with salt and salt water, wrapped it (Ex. 16:4), and announced its arrival to the father (Jer. 20:15). Also they placed a grain of salt in the hollow of a decayed tooth, which was considered a cure for the toothache. Elisha threw salt into the pool of Jericho, to purify it (2 Kings 2:20, 21). Gargling with salt water was also used by the Israelites for treatment of sore throats. Sometimes conquered cities had their fields sown with salt to render them useless for crops for a period of time. The effect of salt on vegetation was to render the land infertile (Deut. 29:23).

Where did the Israelites obtain their salt? Salt was found in many different places such as in salt-pits. Certainly salt was abundant in the salt (Dead) Sea (Num. 34:12, Deut. 3:17) and most of their salt came from the Dead Sea. One hundred pounds of water from the Dead Sea yielded 24.5 pounds of salt. Victories were celebrated in the valley of salt (2 Sam. 8:13, 2 Kings 14:7, 1 Chron. 18:12). The valley of salt can be identified with the plain extending from the southern end of the Dead Sea to the foot of the cliffs that cross the valley from side to side and from the southern margin of the Ghor. Ghor is a region of the Jordan Valley between the Sea of Galilee and the Dead Sea, on the border of Jordan and Israel and the West Bank. Israel and Jordan mainly produced solar salt from the Dead Sea. This method of salt production is where the sea water is placed in shallow pools or pans and allowed to evaporate in the sun. About thirty percent of all the salt produced in the world is produced through solar evaporation procedures. Modern Israel salt industries of course have refined this process.

What is the effect of salt on plants, animals, and microbes? At high concentrations, salt exerts a drying effect on both food and microbes. Salt in water at concentrations of 0.85-0.9% produce an isotonic condition for nonmarine microbes. Because the amount of salt and water are equal on both sides of the cell membrane, water moves across the cell membrane equally in both directions. If microbial cells are suspended in a 5% saline solution, the concentration of water are greater inside the cells than outside (concentrations of water are highest where solute concentrations is lowest). In diffusion, water moves from its area of high concentration to its area of low concentration. Thus, water passes out of the cells at a greater rate than it enters. The result to the cell is plasmolyhsis, which results in growth inhibition and possibly death. Salt also dehydrates foods by drawing out and typing up moisture as it dehydrates microbial cells. Also NaCl ionizes to yield the chlorine ion, which is harmful to organisms. Salt also reduces the solubility of oxygen in the moisture and sensitizes the cell against carbon dioxide.

Of course, proper amount of salt can be added to permit an acid fermentation to take place as has already been discussed. Unfortunately, some microbes are halophiles (salt lovers), which thrive in a 15-30% salt concentration. Spoilage of foods by these microbes is not prevented through use of brine. These microbes will be mentioned later when the Dead Sea is discussed.

Honey. Honey is a sugar substance deposited in the honeycomb by honeybees. Honeycomb is a mass of hexagonal wax cells built by honeybees in their nest to contain brood and stores of honey. Sometimes it was found in the carcasses of dead animals such as the lion (Judges 14:8). Honey was also found in rocks (Deut. 32:13, Ps. 81:16). It was also found on the ground (1 Samuel 14:25) and in the woods (Jer. 41:8). In the region, honey was plentiful in Canaan (Ex. 3:8, Lev. 20:24, Deut 8:8) where the Canaanites, Hittites, Amorites, Perizzites, Hivites and Jebusites lived. Honey was also abundant in Assyria (2 Kings 18:32) and Egypt (Num. 16:13).

Honey was an esteemed food for the Israelites (Prob. 24:13). They ate it with the honeycomb (Song of Solomon 5:1, Luke 24:42), and with milk (Song of Sol. 4:11). They also ate honey with butter (Isa. 7:15,22), with locusts (Matt. 3:4, Mark 1:6) and mixed it with flour (Ex. 16:31, Ezek. 16:13).

Honey was thought to be valuable as it was given as a present (Gen. 43:11, 1 Kings 14:3). Honey was evidently plentiful enough to be exported (Ezek. 27:17).

Honey was regarded as having medicinal properties (Pr. 16:24). Honey was used to treat infected wounds as long ago as 2,000 years before bacteria were discovered as the cause of infection. More recently, honey was reported to have an inhibitory effect to around sixty species of bacteria including aerobes and anaerobes, gram-positive and gram-negatives. An antifungal action of honey has also been observed on some yeasts and molds including the common dermatophytes.

The obvious antimicrobial property of honey is due to its osmotic effect caused by its high sugar content. This also makes water unavailable to organisms essentially in the same manner as salt.

Honey is essentially a highly concentrated water solution of two sugars, dextrose and levulose, with small amounts of at least twenty-two other more complex sugars. Thus, honey is above all is a carbohydrate material with 95-99.9% of the solids being sugar.

Besides honey having a high osmolarity, which kills microbes by drawing water from the cells, honey has a pH value of about four. This acidity inhibits the growth of most bacteria. A third antimicrobial activity of honey is due to hydrogen peroxide. This agent was referred to as "inhibine" prior to its identification as hydrogen peroxide. Hydrogen peroxide is an oxidizing agent. It disrupts disulfide bonds in proteins of cells and this disrupts the structure of membranes and proteins. Hydrogen peroxide also forms the highly reactive oxidizing agent superoxide. When hydrogen peroxide breaks down to oxygen and water, the oxygen kills obligate anaerobes present in wounds. A fourth antimicrobial agent in honey comes from the floral nectar components. These unknown substances could be flavonoids produced by plants, which are found in honey. Plants produce these carotenoid pigments to render the plant resistant to attack by microbes. Honeybees enhance the medicinal value by condensing the nectars to honey thus making the antimicrobial properties more concentrated.

Honey is produced from many different floral sources so its antimicrobial activity would vary; however, today honey could be used in dressing infected wounds as they did in ancient times.

Figs. Fig is the name of a fruit and the plant on which it grows. The first record of the fig in the Bible is in Gen. 3:7, where Adam and Eve are said to be clothed with girdles made from the fig-tree's broad leaves. The fig tree seems to have been an early inhabitant of Palestine like the grape vine and the olive (Judges 9:10-11). Green or winter figs (Rev. 6:13) were on the branches during the winter and became ruddy during early spring, but they were small. Late figs, which came in August, were purplish, green, whitish, or nearly black. Both fresh and dried figs were pressed into cakes (1Sam. 25:18) and were popular as food.

Isaiah prescribed a lump or cake of dried figs as a poultice to be laid on Hezekiah's boil (inflamed ulcer) which was thought by some scholars to be produced by the plague. Figs are considered useful in ripening and soothing inflammatory ulcers and it worked on Hezekiah (2 Kings 20:7, Isa. 28:21). Isaiah said, take a lump of figs and they took and laid it on the boil, and he recovered. Much of the fig's medicinal value is due to its high levels of vitamins A and C, magnesium, flavonoids and benzaldehyde. Since dried figs contain 50% invert sugar and some sucrose, these sugars could act as an osmotic "drawing" agent.

Wine. Wine in the Bible was the product made by the alcoholic fermentation of crushed grapes or grape juice by yeasts and a subsequent aging process. The sugar content of the grapes should be 20-22%. Generally, about 50% of the sugar content is converted into alcohol until the level of alcohol reaches about 17%. At this concentration, the yeast is killed.

Wine in the Bible has great importance both symbolically and as an article of commerce. In this section wine will be discussed when used in medicinal purposes. Wine is advocated in 1 Tim. 5:23 "Drink no longer water, but use a little wine for the stomach's sake and thine often infirmities." Current advice on the benefits of wine were probably anticipated by Paul in his letter to Timothy. The Good Samaritan treated the wounds of the injured man with wine and oil (Luke 10:34).

Certainly the alcohol (ethanol) in wine is a coagulant and denaturizer of cell proteins of microbes. Alcohol can disrupt membranes and dissolve many lipids, including the lipid component of enveloped viruses. Studies have been done determining the antibacterial activity of Chilean red wine against *Helicobacter pylori,* the causative agent of peptic ulcers. The chemical component of wine that showed antibacterial activity was mainly resveratrol. Bonner et al. also found wine consumption reduces the odds of infection with *H. pylori.* Dilutions of red and white wine were found more effective than similar dilutions of bismuth salicylate. Weisse et al. found the antibacterial property of wine is mainly responsible for wine's use as a digestive agent.

Again wine, which was tried and proven in the Bible to be medicinal, is proven in modern time to have antibacterial properties.

Leeks, onions and garlic. Leeks, onions and garlic were mentioned to have been eaten by the Israelites in the Bible. In Numbers 11:5, the children of Israel lament the meager desert diet caused by the Exodus: "We remember the fish, which we did eat in Egypt freely, the cucumbers and the melons and the leeks and the onions and the garlic."

The leek is a versatile vegetable that has not received the same appreciation in the United States as it has in Europe. Leeks have been cultivated since at least 3000 B.C., and they are native to the region stretching from Israel to India. They have been cultivated for so long that their beginnings are uncertain. Phoenician traders introduced the leek to Wales when they engaged in the tin trade in the British Isles. The Welsh certainly knew the medicinal effects of leeks and garlic as stated in this Welsh rhyme: "Eat leeks in March and wild garlic in May, and all the year after the physicians may play." The phytochemical in leeks that appears to be inhibitory to microbes is allium, which is also found in onions and garlic. Thus the Israelites health was benefited by including leeks in their diet.

Because onions are small and their tissues leave little or no trace, there is no conclusive evidence about the place and time of the onion's origin. In Egypt, onions can be traced back to 3500 B.C. King Rameses IV, who died in 1160 B.C., was entombed with onions in his eye sockets. Some Egyptologists think it was because onions were known for their strong antiseptic qualities, which construed as magical, would be handy in the afterlife. Whitaker (1976) lists twenty-eight sulfur compounds in the intact onion. Many are precursors to the substance responsible for the flavor, odor, and antibacterial activities. Enzymes like allinase and lachrymatoryt-factor synthase transform volatile sulphur compounds like sulfoxide to iso allin and propiin in onion. Fresh onion juices have been found germicidal against *Bacillus subtilis, Escherichia coli,* and *Saccharomyces cerevisiae.*

The Israelites used garlic extensively in cooking. This substance has been studied most and is probably the most potent of healing plants mentioned in the Bible. Garlic has been used throughout history for the treatment of a wide variety of conditions. Its usage predates written history. The *Codex Ebers,* an Egyptian medical papyrus dating to about 1550 B.C., mentions garlic as effective remedy for a variety of ailments, including bites, worms, and tumors. Garlic's antimicrobial activity was noted by Louis Pasteur in 1858. Garlic has been shown to have broad-spectrum antimicrobial activity against many genera of bacteria, viruses, worms, and fungi, as summarized in many publications.

These results support the historical use of garlic in the treatment of a variety of microbial conditions and other conditions such as lowering cholesterol levels in the blood. The microbicidal substance acrolein was isolated from garlic and onions by Frazier and Westhoff (1988). In a review by Cowan (1999), allicin and ajoene were the main antimicrobial substances cited to be in garlic. It is not surprising that the Israelites were provided with garlic for various health benefits. This and other spices aided them in maintaining good health.

Cumin. The seed of a small herbaceous plant, the *Nigella sativa,* is widely cultivated in the Mediterranean region of the world. The oldest reference to the cumin plant is in the Old Testament (Isa. 28:25, 27). It is used to flavor dishes, and particularly bread. It resembles the caraway in flavor and appearance. It is said to have medicinal properties. The plant is still beaten out with rods (Isa. 28:27), to preserve the small soft seeds.

In the New Testament the scribes and Pharisees, paid tithes of "mint, anisa and cumin (Matt. 23:23) and were charged by Jesus with neglecting weightier matters of the law.

Historically, black cumin seed was mentioned 1,400 years ago. Black seed was found in the tomb of King Tutankhamun, meaning it held and exalted status among the highest echelons of society in that day.

Black seed has been used for a wide range of ailments and conditions. It has been used to treat bronchial asthma, bronchitis, rheumatism and dysentery. It has also been found to boost the immune system, function as an anticancer agent, and to treat skin conditions, including eczema, abscesses, and boils. The essential oil of black cumin is antimicrobial and aids to rid the intestine of worms. Cumin has been found to kill up to 80% of the food-borne bacteria tested.

Anise. *Pimpinella anisum* of the family *Umbelliferae* is a relative to the carrot family. The seed of anise is often used as flavoring in cooking. The word anise is found only in Matt. 23:23. The plant is native to the Mediterranean region but has been cultivated elsewhere for its aromatic and medicinal qualities such as in Egypt and later by the Greeks. Probably the name anise used in the Bible is dill. In the Middle East, there are many common wild relatives including fennel, which are all used in the same manner. All of these annual plants with yellow flowers possess pungent oil in all parts of the plant. The oil, composed mainly of anethole but also contains estragol, methyl chavicol, furanno-coumarins, flavonoid glycosides and fatty acids is used in medicinals to control infection, perfumes, beverages, pickles, and assists in digestion.

Mint. Peppermint, *Mentha piperita,* is one of the trifles which were tithed, but included also allied plants, such as the horse mint (*M. sylvestris),* which grows wild all over Palestine. In fact, the horse-mint might really be the mint mentioned in Matt. 23:23 and Luke 11:42. Scriptural reference mainly point out the hypocrisy of the Pharisees, who tithed even the commonest garden herbs.

Mint is the name of a whole family of plants with over thirty species of perennial herbs. Mints interbreed so easily it is hard for the experts to distinguish and separate all the varieties. The most common are peppermint and spearmint.

Mint grows in all parts of the world. Both the leaf and the oil are used for flavoring in cooking and in making perfume. Mint is also used in medicine. All mints have the volatile oil menthol, which gives mint the characteristic cooling, cleansing feeling. Menthol is also called peppermint camphor or crystalline organic compound.

Since mint plants were available in Bible times, and it appears the indigenous people of the time were herbalists, they received benefit from mints being antiseptics. They used mints in salads and were used as digestive aids.

Currently, peppermint is one of the most economically important medicinal crops produced in the U.S. Peppermint is on the FDA's GRAS (generally recognized as safe) list and thus can be used extensively for addition to foods without further testing. Peppermint has been proven helpful in the relief of the common cold, which is caused by viruses. Peppermint oil has even been used externally for the treatment of chicken pox, another viral disease.

Mandrakes. Mandrakes are any of six plant species belonging to the genus *Mandragora* belonging to the nightshade family (Solanaceae). They are native to the Mediterranean region and the Himalayas. The best-known species *M. officinoerium* has parsley-shaped roots which are often branched. It has a short stem bearing a tuft of ovate flowers. The flowers are solitary with a purple bell-shaped corolla and the fruit is small orange-colored berries with a strong apple-like scent. One of the first records of the Mandrake dates back to the Bible where the pleasant smell of the plant was recorded in Song of Sol. 7:13. The ancients used the plant as a medicine such as to treat scrofulous tumors and ulcers.

People have long had superstitious beliefs about the mandrakes. According to one superstition, the plant shrieks when it is pulled out of the soil. After the plant had been pulled, it could be used for beneficial purposes, such as healing, inducing love, facilitating pregnancy, and providing soothing sleep. In Gen. 30:14-17 mandrakes were used as a stimulant to conception.

People in the United States and Canada often use the name mandrake for the May apple that should not be confused with the mandrake that is discussed in the Bible.

Mustard. The seed of this plant is used proverbially for anything exceedingly small. Controversy surrounds the identification of the plant whose seed was used by Christ as an illustration of something that develops rapidly from small beginnings, such as the kingdom of heaven (Matt. 13:31; Mk 4:31; Luke 13:19) or the faith of an individual (Matt. 17:20; Luke 17:6).

Some scholars think that the black mustard (*Brassica nigra)* has been inferred since in New Testament times its seeds were cultivated for their oil as well as for culinary purposes. Others have identified the mustard of Christ's parables with white mustard (*Brassica hirta),* a closely related species.

The aroma and flavor of mustard comes from the essential oil contained as glucosides inside the seeds. Powdered mustard has essentially no aroma until it is moistened. The enzymatic action of myrosin on the glucoside sinigrin in black mustard or on sinalbiln in white mustard releases the mustard oil, the compound responsible for the pungency.

For centuries, mustard was considered everyone's cure-all. The Romans introduced it to Britain. Mustard seed was used medicinally by Hippocrates, among other ancient physicians. Mustard plaster was used by the Greek physician Dioscoride for pulmonary congestion. Mustard is a powerful germicide; surgeons used it to disinfect their hands with a paste of mustard seed and water. Mustard plasters have been used in medicine for their counterirritant properties in treating chest colds and other ailments.

Mustard is the most widely used spice in the U.S. aside from black pepper. Black and white mustards are generally recognized as safe (GRAS) for human consumption as spices, natural flavorings and as plant extracts. Mustard was found to be effective against mycotoxigenic *Aspergillus* by Azzouz and Bullerman.

Cinnamon. Cinnamon is a spice used in cooking and in flavoring candies. It is made from the inner bark of branches of the cinnamon laurel tree. The cinnamon tree grows as high as 30 feet (9 meters), and has oval leaves and tiny pale-yellow flowers. Cinnamon trees grown for their bark are usually kept small. This is done by cutting the tree close to the lower buds. The bark of the lower branches is peeled for use as cinnamon. As the bark dries, it curls up and turns light brown.

Cinnamon is a delicious spice well known for its aromatic qualities. Cinnamon was used in the holy oil used in the tabernacle to anoint priests and sacred vessels, as cited in Ex. 30:23-25. Moses used the oil in Prov. 7:17 as a perfume for the bed. In Rev. 18:13 it is mentioned among the merchandise of Babylon.

Cinnamon is still a favorite perfume and flavoring substance in Palestine. References to the medicinal uses of this product by the Israelites is missing, but it is not unreasonable to assume they might have used it in medicine or to prolong shelf-life of their food.

The inhibitory chemicals in cinnamon are cinnamic aldehyde and eugenol. Cinnamon is usually more bacteriostatic than other spices. Recent studies have shown that cinnamon may kill *Escherichia coli* O157:H7, which causes food poisoning in 20,000 Americans each year. Also growth and aflatoxin production by *Aspergillus parasiticus* in broth were inhibited by 200-300 ppm of cinnamon and clove oils, by 150 ppm cinnamic aldehyde and by 125 ppm eugenol (Bullermon et al. 1977).

Hyssop. Hyssop is mentioned several times in Scripture. This plant (Gk. Hyssopos) is difficult to identify, since the references do not always suggest the same species.

It was used for sprinkling blood (Ex. 12:22), and in the ritual of the cleansing of lepers (Lev. 14:4; Num. 19:6). In this purification of lepers, hyssop, with cedar and scarlet wool was used; also for purification for the people with plague (Lev. 14:49-52), and the red heifer sacrifice (Num. 19:2-6; Heb. 9:19). Hyssop was an insignificant plant growing out of the wall (1 Kings 4:33). It could afford a branch strong enough to support a wet sponge (Jer. 19:29).

If hyssop mentioned in the Bible was not the herb *Hyssopus officinalis,* which is not native to Palestine, then what species was it? The plant, which is considered more probable the hyssop of the Mosaic ritual, is a species of *Origanum maru.* Like *H. officinalis* it belongs to the family Lamiaceae, which is noted for aromatic and detergent properties, and can easily be made into an extract for purposes of sprinkling. *O. maru* is also known as Syrian oregano. Oil of oregano has currently been studied as an antimicrobial. The two main inhibitors of Oregano are carvacrol and thymol. They appear to interfere with the respiration of bacteria.

Origanum vulgare - "Greek oregano"

Some ailments that oregano has been used on include allergies, bronchitis, candidiasis, colds, cold sores, diarrhea, the flu, and sinusitis. Also the oil of oregano was found bactericidal for *Listeria monocytogenes, Stapyhylococcus aureus, Escherichia coli, Yersenia enterocolitica and Pseudomonas aeruginosa.*

Rue. Like many Mediterranean plants, Rue has Biblical connections. Rue, also known as Herb-of-grace, (*Ruta graveolens) was* a spice which the Pharisees tithed (Luke 11:42). Throughout history, Rue has been used for everything from flavoring cheese to driving out evil spirits. Rue's silvery leaves are very ornamental in the garden. Rue reaches about two feet when in bloom having little yellow star-crossed flowers. In New Testament times, Rue was prized for its medicinal values, having alleged disinfectant and antiseptic properties. In many parts of the world Rue is taken orally to induce abortion or to initiate menstrual periods. Topically, rue oil may be used to treat arthritis, pains, bruises or sprains. It may also be applied to ward off insects. Folk medicine uses Rue for many ailments, including cramps, diarrhea, earache, toothache, fever, indigestion, intestinal worms, hepatitis, sore throat and skin inflammation. Early research indicates that rutin, a photochemical found in rue, may aid to inhibit tumor formation in the skin.

Aloe. There are five references in the Bible to aloe (John 19:39, Num. 24:6, Ps. 45:8, Prov. 7:17, Song of Sol. 4:14). The medicinal aloes in the family Liliacea is not alluded to. Biblical aloe is probably the modern eagle-wood (*Aquilaria agallechum*). From it was derived a precious spice used in Biblical times for perfuming garments and beds. The fragrant parts are those that are diseased. The odoriferous qualities are due to the infiltration with resin. The development of this change in the wood is hastened by burying it in the ground. The perplexing question of the reference in Numbers 24:6 may suggest that the tree grew in the Jordan valley, but Balaam must have referred to this tree from hearsay or some other tree of the same name that grew in the Jordan valley.

Did the Hebrews know about aloe, of which there are over 500 species? The common name for this plant is Aloe Vera, Curacoa Aloe, Barbados Aloe, and Lilly of the Desert. In about 1750 BC, Sumerian clay tablets indicated the use of Aloe Vera for medical purposes. By 1500 BC, aloe is mentioned in Egypt's Papyrus Embers where formulas containing aloe are described for a variety of illnesses, both external and internal.

Since the Jews lived in Egypt for 400 years, perhaps they learned of Aloe Vera during this time. It is quite probable since Joseph was second in command in Egypt and Moses was raised in the Royal Court.

Coriander. Coriander, *Coriandium sativum*, is a common cultivated plant all over the East. This umbelliferous plant was used as early as 1550 B.C. for culinary and medicinal purposes. Coriander is mentioned in Ex. 16:3 and Num. 11:7. Presently the spice is mainly used for flavoring foods and medicine. The Israelites could have used it medicinally as it has a carminative action on the stomach. There were several microbial diseases that caused them to have a sharp pain in their bowels.

Spices. Spice is associated in the Bible with royalty and wisdom. Spice speaks of extravagance, wealth and splendor (2 Chron. 9:9). It symbolizes the sacramental goodness of the world, the oils and fragrances of love (Song of Sol. 5:1), the beauty of an oriental garden with its blossoms and beds of aromatic herbs (Song of Sol. 6:2). Spice, frankincense and myrrh represent a life comfortably rooted in the land, partaking of the wealth of the king.

Spice caravans pioneered the trading routes from Northern India to Sumeria, Akkod, and Egypt at a very early period, and subsequently these roots became an important factor in cultural exchanges. While many spices were brought to Palestine from Mesopotamia and India, a number of those in common use were the product of the country itself. In Old Testament times the Palestinian spice trade was carefully protected. Solomon derived considerable revenue by charging tolls of the caravans passing through his region.

Since the spice trade route went through Israel, many spices were available and used. For whatever reason, many of these spices that had medicinal uses were not recorded in the Bible.

Additional Reading on leeks, onions and garlic:

Cowan, M.M. *Plant Products as Antimicrobial Agents*. Clin. Microbiol. Rev. *12*:564-582, 1999.

Frazier, W.C. and D.C. Westhoff. *Food Microbiology, 4th Ed.*, McGraw-Hill Book Co., New York, 1988.

Tyler, V. E. *The New Honest Herbal.* George F. Stickly Co., Philadelphia, PA, 1987.

Whitaker, J.R. *Development of Flavor, Odor and Pungency in Onion and Garlic.* Advances in Food Res. 22:73-133, 1976.

http://www.equinecentre.com.au/health.nutraceuticals.garlic.shtml

Additional Reading on cumin:

http://www.theramune.com/blackseed.html

Additional Reading on mint:

Blumenthal, M. *The Complete German Commission E Monographs: Therapeutic Guide to Herbal Medicine.* Austin: American Botanical Council, 1998.

Additional Reading on mandrakes:

http://www.monstrous.com/monsters/mandrake.htm

Additional Reading on mustard:

Azzouz, M.A., and L.R. Bullerman. *Comparative Antimycotic Effects of Selected Herbs and Spices, Plant Components and Commercial Antifungal Agents.* J. Food Protect. 45:1248-130l, 1982.

Additional Reading on cinnamon:

Bullerman, L.B., F.Y. Lieu, and S.A. Seren. *Inhibition of Growth and Aflatoxin Production by Cinnamon and Clove Oils, Cinnamic Aldehyde and Eugenol.* J. Food Sci. 42d:1107-1109, 1116, 1977.

Additional Reading about hyssop:

http://www.mountainvalleygrowers.com/orimaru.htm

EMBALMING

Embalming is a process to preserve human remains to forestall decomposition, and in Bible times was done by aromatics (Gen. 50:2, 3, 26). Perhaps the Old World culture that had developed embalming to the greatest extent was that of Egypt, who developed the process of mummification as early as 4000 B.C. They believed their God of the dead, Anubis, was there inventor of embalming, and that preservation of the body empowered the soul in future reunion with the body.

Though the Bible (Gen. 50:23-26) states that Jacob and Joseph were embalmed, the latter by his servants and physicians, this procedure was not used much. Embalming was hardly practiced among the Hebrews, as evidenced by the conditions of the bodies found in thousands of Hebrew tombs. They rarely embalmed because of their religious beliefs, their opposition to the Egyptian religion, and because of its high cost to the typically poor Israelite.

The use of spices in the burial of King Asa (2 Chron. 16:14) and Jesus (John 19:39-40) was not to embalm but to purify them ceremonially. Myrrh was used by the ancients as a perfume and for embalming. It is a gum common in Arabia, Egypt and Abyssinia. Certainly the spices, having antimicrobial activity, slowed down body decomposition.

Embalming and mourning usually took seventy days, but Jacob's took forty (Gen. 50:1-3).

Physicians & Medicine

Among the ancients, Egyptian medicine is the best known, though it was mainly magical in nature. Their early physicians were often temple priests organized under a chief who dealt principally with hygiene and diet. The Egyptians claimed to invention of the healing art, and their "many medicines" are mentioned (Jer. 46:11).

Doctors were so highly esteemed in Egypt that sometimes they were deified. Egyptian doctors served as embalmers (Gen. 50:12) and as apothecaries, preparing tonics, hair pomades, and other cosmetics.

In the Bible there is the concept that illness and plagues are punishment for sin (Ex. 15:25-26; Lev. 26:14-16; Deut. 7:12-16) or due to Satan (Job 1). Yet, the physician was regarded as a messenger of God (Deut. 32:39). While Israel regarded the Lord as the physician of Israel (Ex. 15:26), human practitioners were approved (Ex. 21:18-19). Joseph hired house physicians (Gen. 50:2), and Isaiah refers to a surgeon or wound dresser (Isa. 3:7).

Occasionally there were prophets or holy men, who engaged in diagnostic or healing activities. Elijah revived a dead child (1 Kings 17:17-22), and his disciple Elisha performed a similar act (2 Kings 4:18-20, 34-35). A prophet restored the paralyzed hand of King Jeroboam (1 Kings 13:4-6). Isaiah cured King Hezekiah of an inflammation by applying a poultice made of figs (2 Kings 20:7).

While the word physician is rarely used in the Bible, it denotes substantially the same meaning as our present-day medical practitioner (Ex. 15:26; Jer. 8:22; Luke. 8:43).

The physician's duties are not clearly defined in the Bible. One can infer what they may have been. Prv. 30:15 may show knowledge of the medicinal use of leeches; but this inference cannot be drawn with certainty from the context. In Ezek. 30:21 there is references to a primitive form of bone setting. Surgical operations mentioned in the Bible are circumcision and castration (Deut. 23:1). There is recognition of inflammation and abscesses (Deut. 28:35, gangrene and discharges (Ps. 38:5; Prov. 12:4, 14:30). Sores are treated with local application (Isa. 1:6; Jer. 8:22, 51:8). The Good Samaritan used wine and oil as local treatment (Luke. 10:34). Wine is also mentioned as a remedy and stimulant (Prov. 31:6; 1 Tim. 5:23). The value of therapeutic baths is mentioned in 2 Kings 5:10.

Attitudes toward physicians appear to have varied. Asa is denounced for having consulted physicians (2 Chron. 16:12). Job chides his comforters "as physicians of no value" (Job 13:4). Jesus, considered the physician of the soul (Matt. 9:12), mentions physicians twice in the New Testament (Luke. 4:23, 5:31). Luke is described by Paul as the beloved physician (Col. 4:14). Then there is the woman "who had suffered many things of many physicians" (Mark 5:26-27). Then she touched the garment of the Great Physician and was healed. Of course, this healing power of Jesus is available to anyone that will accept him as their personal savior.

SANITARY CODE

The Jews of Biblical times were responsible for the first written public health laws. This sanitary code spared them from many a plague.

The Jews quarantined for contagious diseases such as leprosy (Lev. 13; Num. 5:2-3). They were also given the rites and sacrifices in cleansing of the leper (Lev. 14:1-32; Deut. 24:8 and disinfecting of houses after contagious diseases (Lev. 14:33, 57).

The Mosaic Law required the burial of excreta (Deut. 23:13-14). The Law pointed out the uncleanness of men and women by their discharges and told them how to cleanse the issues. The Law also told how to purify a woman after childbirth (Lev. 12:1-8; Luke 2:22). In modern times, the New York State Department of Health became alarmed after learning that infectious microbes could spread so quickly by a carrier who failed to wash his hands carefully. In 1960 the Department issued a book describing a method of washing the hands, and the procedure closely approximated the method given in Num. 19.

The Jews were required to circumcise all male children on the eighth day after birth (Gen. 17:10-14; 21:4; Ex. 12:48; Lev. 12:3; Luke 1:59; 2:21; John 7:22-24; Acts 7:8). Advantages of circumcision could include improved cleanliness, less risk of cancer in the penis, less risk of inflammation of the penis and foreskin, less risk of urinary tract infections and possibly less risk of transmitting diseases (including human immunodeficiency virus and cervical cancer).

Moses' writing reveals knowledge of deadly microbes associated with dead or unclean things (Lev. 5:2; 11:24-40; 21:1-4; Num. 5:2; 9:6, 10; 19:11-22; 31:19). The writings also revealed how to inspect and select which food to eat. The Jews as a nation might not have survived their time in the wilderness without the sanitary code. This code implies a knowledge that the Jews did not possess beforehand, and during the times of the Exodus and the wilderness wanderings they scarcely could have discovered for themselves, such as prohibition of foods such as pigs and animals that had died natural deaths, and the burial of excreta, etc.

Additional Reading:

http://www.biblestudy.org/basicart/peruncin

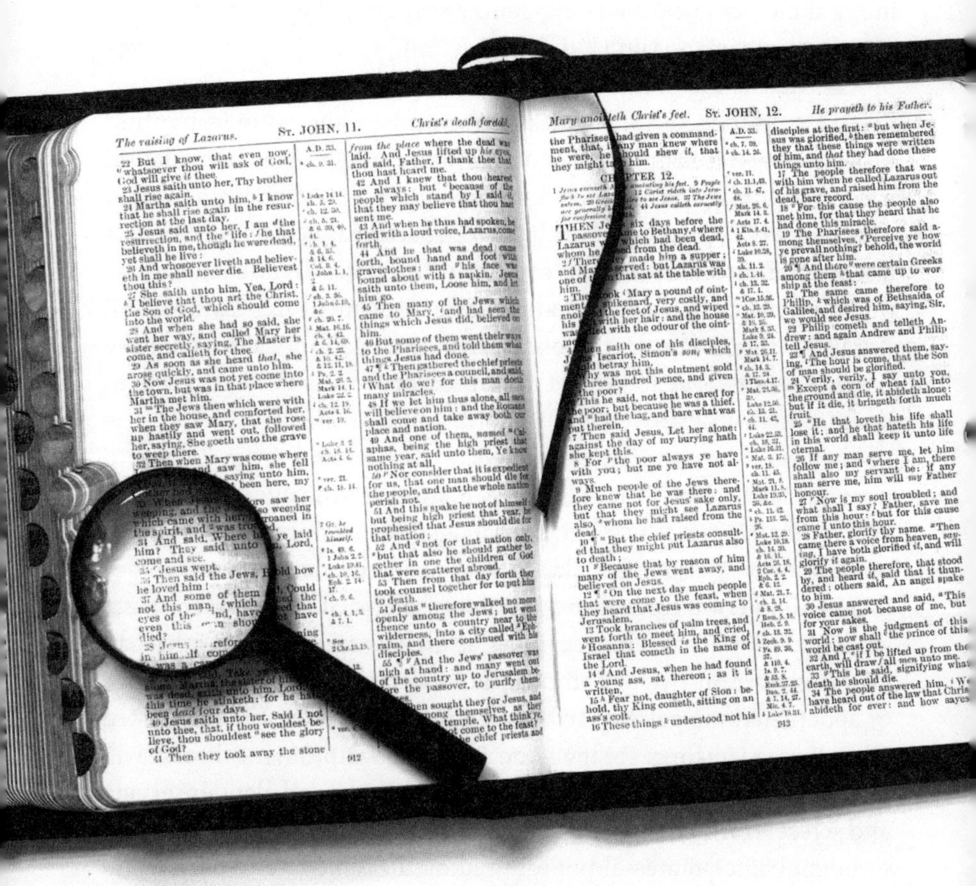

Is the Dead Sea Dead?

The original Dead Sea was an inland lake forty-eight miles long and three to nine miles wide, which receive the waters of the Jordan. Today the sea is about thirty miles long and has receded to two-thirds its original size. The river's flow has been reduced by diversion for households and farms. The surface of the water is on average 1,280 feet below sea level, being the lowest body of water on the surface of the earth. It has no outlet, and the rate of evaporation is so great (temperature reaches 110° F in summer) that the inflow of a few streams and the Jordan serve only to keep the sea level constant. The annual rainfall is about two inches.

The concentrated chemical deposits (salt, potash, magnesium, and calcium chlorides and bromide, 25% of the water) make it a hypersaline basin. The pH of the water is 6.1.

The modern name is of late origin and refers to the apparent total absence of life in its waters. In the middle ages, travelers reported that no birds flew over the Dead Sea, as the air there was poisonous. Gases rising from the water give it an unpleasant odor. Some persons believe the water can cure certain diseases. It has no Scripture warrant: Hebrew writers speak of it as the "Salt Sea" (Gen. 14:3; Num. 34:3), the sea of the Arabah (Deut. 3:17; 4:19) and the east or eastern sea (Ezek. 47:18). Salt was obtained from the shore (Ezek. 47:11). Throughout the Old Testament period the sea acted as a barrier between Judah and Moab and Edom (2 Chron. 20:1-30). It may have been used by small trading boats, as it was in Roman times.

Despite what may seem like harsh conditions, there can be a highly productive ecosystem in the Dead Sea. Extremely halophilic Archaea are a diverse group of procaryotes that can live in hypersaline environments like the Dead Sea. Using 15S ribosomal ribonucleic acid (rRNA) sequencing and other criteria have defined ten genera of extreme halophilic Archaea. Of these ten genera three have been isolated from the Dead Sea: *Halorubrum, Halobaculum,* and *Haloferax.* Other well-adapted and widely distributed extreme halophilic microbes include cyanobacteria, such as *Aphanothec halophytican,* and the green algae *Dunaliella salina.* The eucaryote alga *D. salina* is the major, if not sole oxygenic phototroph in most salt lakes. Organic matter originating from primary production by oxygenic or anoxygenic phototrophs then allows for development of the extremely halophilic Archae

Extreme halophilic Archae stain gram negative, reproduce by binary fission, and do not form resting stages or endospores. Most halobacteria are nonmotile but a few strains are weakly motile by flagellar means.

Thus in modern time we know the Dead Sea is not dead but has a productive ecosystem particularly containing microbes that have adapted to extremophilic, harsh, hypersaline conditions.

Additional Reading:

Hammer, J. *The Dying of the Dead Sea.* Smithsonian. October p. 58-70, 2005.

MANNA

This was the main food given to the Israelites during their forty years of wandering in the wilderness (Ex. 16:4, Num. 11:6). It looked like small, round yellowish-white flakes and tasted like wafers made with honey. It could be ground and used in cooking and baking. It rained from heaven each morning. It was gathered early because it melted in the sun. The Israelites were to collect an amar (about 6 pints, or 2.8 liters) for each person daily. Usually it would not keep overnight, but spoiled because of microbial decomposition and maggots and became malodorous. Twice the usual amount fell on the sixth day to last them over the sabbath. This was preserved by being cooked or baked beforehand (Ex. 16:23-30). The provision of manna did not cease until the Israelites crossed into Cannan, the Promised Land (Josh. 5:12).

Some scholars think manna was secretions by certain insects, which was like gluey sugar produced on the tamarisk shrub. Some identified manna with the juice of the flowering ash, which exudes a "manna" (used in medicine). These phenomenon do not satisfy Biblical data such as the continuity, quantity and six-day periodicity. Even so, God might have used this type of physical basis in providing this food for the Israelites.

Other mention of manna was that an omer of manna was laid up on the ark in a golden pot by Aaron so it could be a witness for future generations (Ex. 16:32-36, Heb. 9:4). It is miraculous that the manna was preserved in the pot. This may be one of the first examples of oligodynamic action, which is the ability of extremely small amounts of a heavy metal to exert a lethal effect upon bacteria. The word comes from two Greek words oligos, meaning small, and dynamics meaning power.

Additional Reading:

http://pediatrics,aappublications.org/cgi/content/abstract/2/3/272

http://www.newadvent.org/cathen/09604a.htm

http://enwikipedia.org/wiki/manna

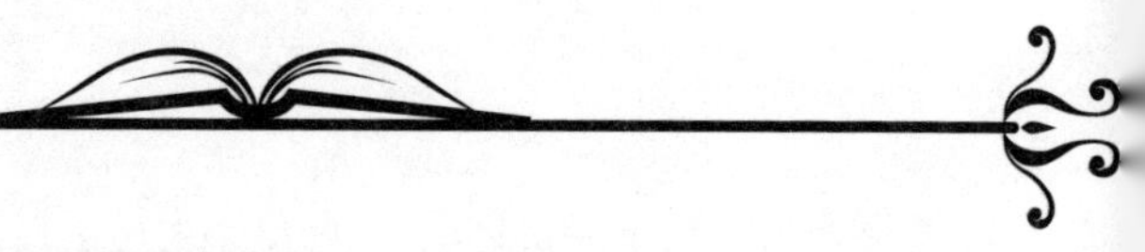

Normal (Indigenous) Microflora of Humans

When Moses led the great multitude of Israelites out of Egypt, God gave him a remarkable promise in Ex. 15:26. "If thou wilt diligently harken to the voice of the Lord thy God, and wilt do that which is right in his sight, and wilt give ear to his commandments, and keep all his statutes, I will put none of these diseases upon thee, which I have brought upon the Egyptians, for I am the Lord that healeth thee."

The Israelites also had good health because they ate a pure food source such as manna. What is not obvious to the non-microbiologists is that all people have normal microbiota of the human body, which aid in good health. The mixture of microbes regularly found at any anatomical site is referred to as the normal flora.

The normal flora of humans is complex and consists of more than four hundred species of bacteria. The complexity of the normal flora depends upon factors such as genetics, age, sex, stress, nutrition and diet of the individual. The normal or indigenous flora of humans consists of a few eucaryotic fungi and protists, and some methanogenic archaea that inhabit the large intestinal tract, but bacteria are in largest numbers and will be emphasized in this chapter. Viruses and parasites are not considered normal flora because they are not commensals and do not aid the host.

The human body contains about ten trillion eucaryotic cells. It harbors about one hundred trillion procaryotic and eucaryotic microbes with the bacteria dominating.

We are continually exposed to microbes and many have established residence in and on our bodies. Most do little or no harm, and some are beneficial. We will examine some of the microbes that inhabit the healthy human adult body under normal circumstances and summarize the benefits of these indigenous microbes.

The skin is the largest and most accessible of all organs. Although many microbes come into contact with the skin surface, this relatively hostile environment does not support the growth of most organisms. The normal skin residents consist primarily of *Staphylococcus, Corynebacterium, Propionibacterium* and yeasts. Gram-negative bacilli do not permanently colonize the skin surfaces (with the exception of *Acinetobacter*) because the skin is too dry.

The flora of the mouth is unique in that it is among the most diverse and abundant of the body. Microhabitats including the cheek epithelium, gingiva, tongue, floor of the mouth, and tooth enamel, provide numerous adaptive niches for abundant different species to colonize. The most common residents are Streptococci such as *Streptococcus sanguis, S. salivarius* and *S. mitis* that colonize the smooth superficial epithelial surfaces. Saliva normally has a high bacteria count that tends to make a human bite dangerous.

The most common microbes found to colonize the outer ear is coagulase-negative *Staphylococcus.* Other microbes colonizing the skin have been isolated from this site.

The surface of the eye is colonized by coagulase-negative staphylococci, as well as rare microbes found in the nasopharynx such as *Haemophilus* spp., *Neisseria* spp. and viridans streptococci.

The first microbes to colonize the upper respiratory tract (nasal passages and pharynx) are predominantly oral streptococci. *Staphylococcus aureus* preferentially resides in the nasal entrance, nasal vestibule and anterior nasopharynx in about thirty percent of the population. *Neisseria* spp. reside in the mucous membranes of the nasopharynx behind the soft palate. Lower still are assorted streptococci spp. and *Haemophiluls* spp. that colonize the tonsils and lower pharynx. Continuing lower in the respiratory tree (bronchi and lungs are unfavorable habitats for permanent residents.

The human gastrointestinal tract consists of the stomach, small intestine, and large intestine. The stomach is generally colonized with small numbers of acid-tolerant microbes such as the lactic acid bacteria, lactobacilli and streptococci.

The small intestine is colonized with many different microbes.

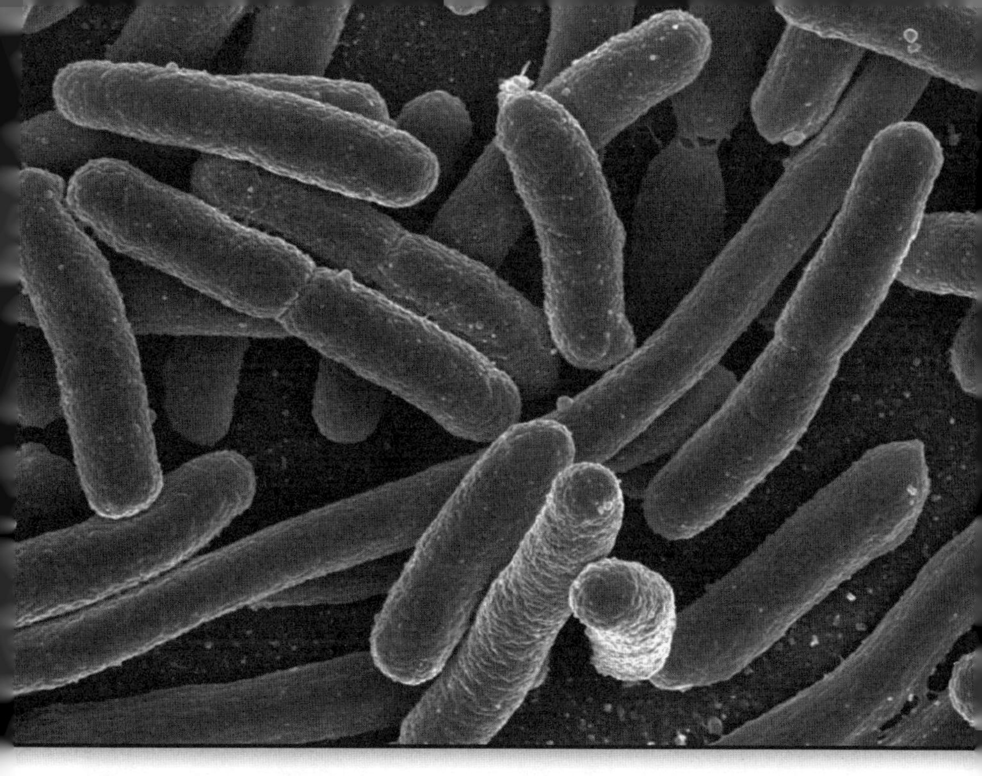

Escherichia coli, grown in culture and adhered to a cover slip. Photo by Rocky Mountain Laboratories, NIAID.

Most of these organisms are anaerobic bacteria. The largest number is in the lower part called the ileum containing 10^5 to 10^7 Bacteria/gram of feces.

More microbes are present in the large intestine than anywhere else in the human body. It has been estimated that more than 10″ bacteria per gram of feces can be found, with anaerobic bacteria in excess by a thousand fold. The activities of falcultative anaerobes such as *Escherichia coli* and *Enterococcus faecalis* consume any oxygen present, making the environs of the large intestine strictly anaerobic. This creates favorable conditions for the growth of obligate anaerobes. Obligate anaerobes isolated include species of *Bacteroides, Clostridium, Bifidobacterium, Fusobacterium* and *Peptostreptococcus.* Various species of *Bacteroides* account for the majority of intestinal anaerobes. So abundant and prolific are these microbes that they constitute 10-30% of the fecal volume.

In general, the anterior urethra and vagina are the only anatomical areas of the genitourinary system that are colonized with microbes. Although the urinary bladder can be transiently colonized with bacteria migrating upstream from the urethra, these should be rapidly cleared by the bactericidal activity of the uroepithelial cells and the flushing action of voided urine.

The primary indigenous populations of the urethra consists of non-hemolytic streptococci, lactobacilli, coagulase-negative staphylococci, corynebacteria, and occasionally coliforms.

The microbial population of the vagina is dramatically influenced by hormonal factors. An important hormone influencing this change in women is the hormone estrogen. New-borne girls are colonized with lactobacilli at the time of birth, and the bacteria predominate for approximately six weeks. After that time, the levels of maternal estrogen have declined and the vaginal flora change to include staphylococci, streptococci, corynebacteria, and Enterobacteriaceae. When estrogen production is initiated at puberty, the microbial flora changes. Lactobacilli (*Lactobacillus acidophilus*) re-emerge as the predominant bacterium, and many other microbes are also isolated, including *Staphylococcus epidermidis, Enterococcus faecalis* and some alpha-hemolyhtic streptococci. From puberty to menopause, the vaginal epithelium contains glycogen due to the action of circulating estrogens. *L. acidophilus* predominates, being able to metabolize glycogen to lactic acid. The lactic acid and other products of metabolism inhibit colonization by all except *L. acidophilus* and a select number of lactic acid bacteria.

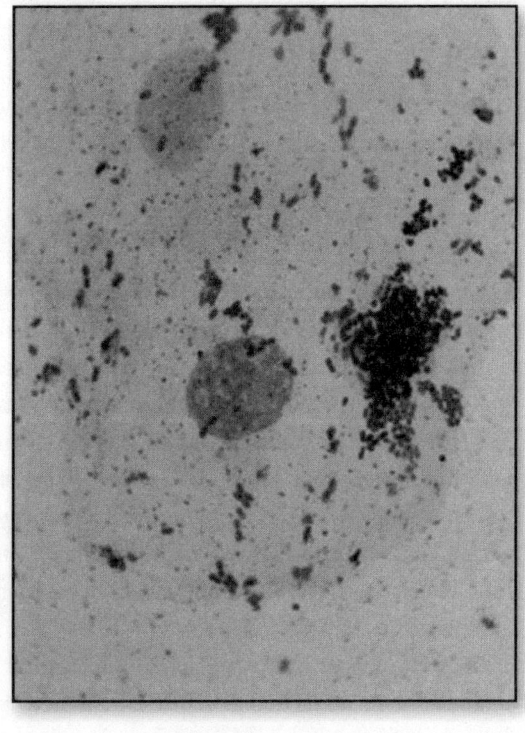

Lactobacillus acidophilus

The resulting low pH of the vaginal epithelium prevents establishment of most bacteria as well as potentially pathogenic yeast. This is a striking example of the protective effect of the normal flora for the human host. How else is the normal flora of humans beneficial to humans as it has been since the beginning of time?

The normal flora stimulates our immune system. Our repertoire of immunoglobulins reflects, in part, the antigenic stimulation by the normal flora. In general, we do not have high antibody titers to the individual fungi, bacteria or viruses that inhabit our body. Nevertheless, even in low concentrations, the antibodies serve as a defense mechanism. Here, then, is a definite benefit from our normal flora. Among the antibodies produced in response to indigenous microbial stimulation are those of the IgA class, which are secreted through mucous membranes.

The normal flora keeps out invaders of our body. In some parts of the body, the normal flora keeps out pathogens. This occurs in several ways. Indigenous bacteria have the physical advantages of previous occupancy, especially on epithelial surfaces. Some commensal bacteria produce substances that are inhibitors to newcomers, such as antibiotics or proteins called bacteriocins. Many enteric species are capable of

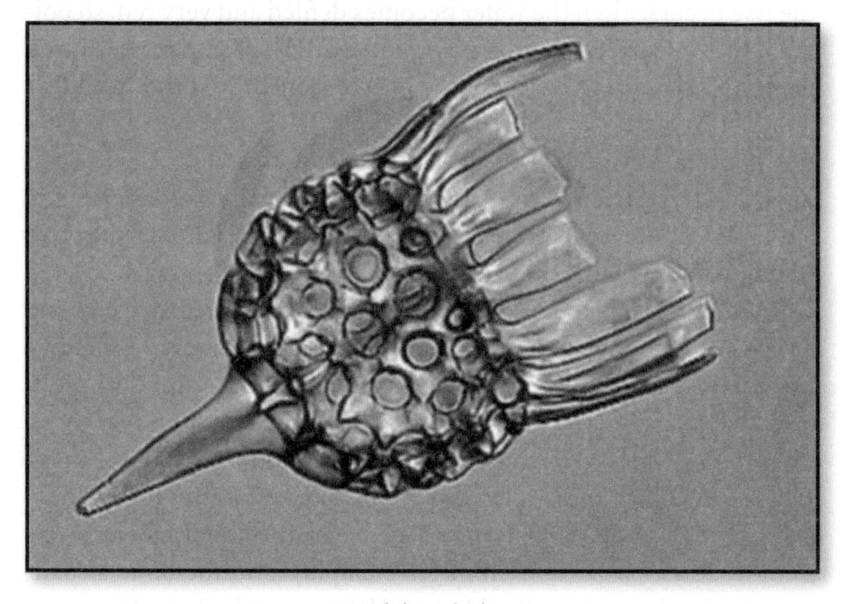

Radiolarian fossil

producing bacteriocins that are bactericidal to related species of but not to themselves. It is not surprising, therefore, that colonization by a new species or a new strain is not a frequent event.

The normal flora also plays a role in human nutrition and metabolism. It is known that several enteric bacteria secrete in excess of their own needs Vitamin K and Vitamin B12, and lactic acid bacteria produce certain, B-vitamins. Also the microbial biomass within us plays a role in recycling certain important compounds such as bile salts. The normal flora also prevents colonization by pathogens by competing for essential nutrients.

Obviously the Israelites lived in a microbial-laden world, which provided them with an excellent normal flora. This normal microbial flora aided them from being diseased.

Additional Reading:

http://www.mansfield.ohio-state.edu/~sabeclon/biol2035.htm

Commiff, R-2013. The body eclectic.

Smithsonian, May. p. 40-47.

OIL

Are you wondering why "Oil" is being explored in this book? The Israeli Jews have a joke that says Moses took a wrong turn because he led the people to the one place in the Middle East with no oil.

I became aware that oil and gas might be in Israel by reading about John Brown. He is a Christian who is using passages from the Bible that tell about verses where God gives blessings to Israel. Using the Bible as a treasure map and hiring good geologists, he founded the Zion Oil and Gas Company. He said, "it's geology confirming the theology."

There are several clues in the Bible that are support for Israel oil. Specifically, Brown cites Deuteronomy where Moses blesses the children of Israel before his death. In Deut. 33:13 it states: "And of Joseph he said, blessed of the Lord be his land, for the precious things of heaven, for the dew, and for the deep that coucheth (lies) beneath." In Deut. 33:19 it states: "They shall call the people unto the mountain, there they shall offer sacrifices of righteousness; for they shall suck of the abundance of the seas, and of treasures hid in the sand." It states, that in the seas and deserts one should find gas and oil. In November of 2009 another company, Delek-Noble, announced a natural gas discovery off Israel's northern coast that will supply Israel's need for fifteen years. The energy exploration experts state that where you find gas, you find oil and vice-versa.

In Deut. 33:24 it states: "And of Asher he said, let Asher be blessed with children; let him be acceptable to his brethren, and let him dip his foot in oil." Here the Hebrew word "shemen" (oil) is used, not olive oil. In Gen. 49:25, Jacob tells his son Joseph that God will give him the blessings of the deep that lieth beneath. Another clue is found in Gen. 14:10, where it is stated that the Dead Sea area was full of slime pits (tar pits). With his Bible and faith, John Brown's company holds 162,000 acres under license in Northern Israel where they expect to find oil and gas under Armageddon. Looking back at the classical book "Epicenter," by Joel C. Rosenberg, he includes several Old Testament scriptures that describe God's promises for Israel to obtain enormous wealth and treasures during end times.

Besides The Zion and Gas Co., there are several other prominent born-again Christian oilmen who are drilling in Israel based on science and their interpretation of some of the same biblical passages. There is a Jewish-owned drilling company that uses the same scriptures for inspiration in finding oil in Israel. Tovis Lusken, a Russian Jew who is a petroleum geologist, owns Givot Olam Exploration Limited Partnership south of Zion Oil. They believe there is a billion barrels of oil in their licensed area from exploratory wells dug in 2004. What is exciting in God's plan is Israel's oil lies west of the 1949 Armistice Line.

Now that it is established there is oil and gas under Israel, some emphasis will be given to some of the microbes contribution to the formation of crude oil. Petroleum not only includes crude oil, but also includes all liquid, gases, and solid hydrocarbons. Under surface pressure and temperature conditions, lighter hydrocarbons such as methane, propane, ethane, and butane occur as gases, but pentane and heavier ones are liquid or solids. Crude oil varies greatly in appearance depending on the composition. It is usually black or dark brown. Petroleum is formed when large quantities of dead organisms, usually zooplankton and algae, are buried underneath sedimentary rock and undergo intense heat and pressure. This surely could have taken place after the Great Flood that covered the whole earth.

Since life certainly was concentrated in the Mesopotamian area, much death occurred there including humans and thus you would expect more oil produced in the region. Over time, hundreds of feet of mud containing organisms accumulated. Anaerobic bacteria removed most of the oxygen, nitrogen, phosphorous, and sulfur, leaving some lipid molecules that tended to be less desirable for the bacteria to decompose. These bacteria and materials fell to the bottom and became the major compounds buried under layers of sediment that form petroleum. When the depth of buried material reached approximately 10,000 feet, heat and pressure through time turned the organisms into different kinds of petroleum. Higher temperatures produced lighter petroleum and lower temperature produced a thick material like asphalt. As the heat continued to alter the substance, gas was produced. The mud and silt become more compressed and turned into a rock known as shale.

There is also evidence of microbial production of hydrocarbons that occurs in certain green algae. An example is the colonial alga *Botryococcus braunii*. Its growth is accompanied by the excretion of long-chain hydrocarbons that have the consistency of oil. In *B. braunii*, approximately 30% of the cell dry weight is petroleum.

In surveying for oil fields, scientists have noted two types of protozoa, the foraminiferans and radiolarians are of special interest to them. These protozoa were a dominant species during formation of the oil fields, and their fossils serve as markers for oil-bearing layers of rock.

If this is not enough oil for Israel, the land also contains oil shale. The Israel Energy Initiative now hopes to extract oil from shale near the Shfela Basin, southwest of Jerusalem. It is predicted Israeli Shale may hold 250 billion barrels of oil. As noted in Deut. 33:19, the shale oil is surely a treasure hid in the sand. Israel's offshore natural gas wells are the "abundance of the seas."

Oil-shales are fine-grained sedimentary rock containing significant amounts of kerogen. Kerogen is an insoluble material formed by the degradation of organic matter and is the main ingredient of hydrocarbons. Oil-shale formation takes place in a number of depositional settings. Oil-shale also has compositional variation, much of which is of algal origin but can include vascular land plants. There is evidence that oil in certain types of oil shale formation originated from green algae like *B. braunii* that grew in ancient lake-beds.

Israel has developed and is testing a method of extracting oil from shale underground. This new method produces oil and water for approximately $35 to $40 a barrel and without harming the environment. Thus it is apparent that companies drilling for oil and gas in Israel using Bible predictions will cause Israel to be an energy superpower in the near future. This will bring peace and wealth both are predicted in the Bible for Israel near end times.

Additional Reading:

http://www.nzpam.govt.nz/cms/petroleum/overview/what-is-petroleum and%20how%20is%20it%20formed

Rosenberg, J.C. 2008. Epicenter: why the current rumblings in the Middle East will change your future. Tyndale House Publishing, Inc., Carol Stream, Illinois

Conclusion

In conclusion, one can say the Bible is not a science book, but definitely is scientifically correct. In searching the Bible, many other interesting facts may be found concerning microbes and topics related to microbiology that are not presented in this book.

For example, in Gen. 7:11, 8:2 and Job 38:16 the springs of the sea are mentioned, which may refer to deep sea hydrothermal vents. Currently two types of vents have been discovered: warm vents that emit fluid from 6-23° C and hot vents, called black smokers, which emit fluid from the deep sea at 270-380° C. At this remote high-pressure environment, scientist initially thought there would be no signs of life. We now know that life there is diverse among these extreme conditions. Science has uncovered hyperthermophilic Archae and an abundant group of invertebrate animals including tubeworms that may reach two meters in length. This unique ecosystem contains high numbers of mussels and giant clams feeding on the chemolithotrophic bacteria springing from the hydrothermal vents.

Another habitat for bacteria mentioned in the Bible include ruminant animals, which posses a rumen where the digestion of cellulose takes place by microorganisms. Common domestic ruminants mentioned in the Bible are cattle (1 Sam. 6:7, Ps. 68:30), sheep (Gen. 32:14, 2 Sam. 12:1-4), and goats (Gen. 27:9, 14, 17). The rumen contains high numbers of bacteria, which also make vitamins and amino acids, which are essential nutrients for the ruminants.

Many plants are mentioned in the Bible. One type used for food were bean plants (Ez. 4:9, 2 Sam. 17:28). Beans are legumes that bear seeds in pods. Bacteria can infect the legumes roots and produce a symbiotic relationship. This symbiotic relationship forms root nodules where nitrogen-fixation takes place and causes the plants to grow better. Many of these root nodule bacteria are found in Israeli soil.

Israel had abundant forests in Biblical times. One tree that is referenced several times in the Bible was the oak (Ez. 27:6, Ez. 6:13, Gen. 35:8, Amos 2:9 and Hos. 4:13). Oaks can form a symbiotic relationship between its roots and fungi termed mycorrhizal. The fungi form mainly sheaths around the roots with little penetration of the root. The mycorrhizae will help plants absorb nutrients and thus grow better, particularly in poor soils.

Pestilence is a complex subject mentioned in the Bible. Pestilence can be defined as an infectious disease that is widespread. Pestilence occurred many times to Israel as punishment for their sins. An example is found in Lev. 26:14-25.

We find described in Rev. 6:7-8 the pale horses of pestilence, and Death. These are signs that precede end-times. Much speculation could be written concerning these infectious microbes that will and currently are causing death. One might infer we are close to the end of times by the number of diseases that have emerged during our lifetime. At the end of the 1970s, I thought possibly my Pathogenic Microbiology class would become obsolete since many diseases were being eradicated. Then in 1981 acquired immunodeficiency syndrome (AIDS) was recognized as a distinct disease. Worldwide, more than 50 million people have been infected with the human immunodeficiency virus (HIV) causing AIDS. *Mycobacterium tuberculosis,* the causative agent of tuberculosis (TB), is rapidly becoming resistant to antibiotics and many cases of TB in the U.S. is occurring in AIDS patients.

Emerging and reemerging infectious diseases that are occurring and will occur during the end of time will be the scope of another book.

Hopefully, people realize that a belief in God does not disqualify a person as a qualified scientist. A scientist whom I admire very much is Sir Isaac Newton (1642-1727). He was an important person or perhaps the most important in formulating modern science. I find it extremely interesting that he was a strong believer in both God and the Bible.

In closing, all that read this book must understand and believe that "Blessed is the Nation whose God is The Lord," (Ps. 33:12).

GLOSSARY

A

Aerobic: An organism that grows in the presence of oxygen.

Algae: Photosynthetic eukaryotic organisms in the kingdom Protista and Plantae.

Allium: There are over 600 members of the Allium family and they have remarkable medicinal powers formed by their enzyme allinase, which act on precursors to form thiosulfinates.

Alpha-hemolytic: These microbes produce an alpha hemolysin. On blood agar plates alpha hemolysin partially lyse (rupture) the red blood cells, which produce a green color around the cell colonies. The hemolytic reactions on blood agar are particularly used for the preliminary identification of streptococci.

Amebiasis: It is an infection with the intestinal protozoan *Entamoeba histolytica.*

Anaerobic: A microbe that has to live in the absence of oxygen and is killed in its presence.

Anoxygenic: Carrying out photosynthesis in which oxygen is not produced as a by-product. There are several groups of bacteria that carry on anoxygenic photosynthesis such as the green sulfur bacteria, purple non-sulfur bacteria and acidobacteria.

Antimicrobics: Any chemotherapeutic agents used to treat diseases caused by microorganisms.

Antiseptic: A chemical agent that can be used on tissues to kill microbes or to inhibit their growth.

Archaea: They constitute a domain of single-celled microbes. These microorganisms contain no true nucleus or any other membrane-bound organelles in their cells.

Ascaris suum: A large roundworm (a parasitic nematode) of pigs.

Ascospore: A sexual spore produced in a sac-like cell (ascus) of a sac fungus (Ascomycetes).

Ataxia: An unsteady gait or inability to walk, a lack of muscle coordination. In the tertiary stage of syphilis neurological damage can occur which includes thickening of the meninges and ataxia.

Atrophy: Refers to a wasting away or loss of muscle tissue due to disease, injury or lack of use.

B

Bacilli: A rod shaped bacteria.

Bacteremia: An infection where bacteria are transported in the blood but do not multiply in the blood.

Bactericidal: Any agent that kills bacteria.

Bacteriostatic: Referring to an agent that inhibits the growth of bacteria but does not kill them.

Bolus: A round mass: a large quantity of a substance.

Bright-field Microscope: A microscope in which the specimen is viewed under visible light. The illumination is produced by the passage of visible light through a condenser.

Budding: Reproductive mechanism that particularly occurs in yeast where a small new cell develops from the surface of an existing mother cell.

C

Candidiasis: It is a yeast infection caused by Candida albicans that causes thrush (in the mouth) or vaginitis.

Capsule: A protective structure secreted outside the cell wall of a microbe composed mainly of polysaccharides.

Cell-mediated immune response: The immune response brought about by the direct action of T cells, such as T cytotoxic, T suppressor and T helper to destroy microbe-infected cells, tumor cells or organ transplants.

Cellulitis: A bacterial infection of the dermis, which is the deep layer of skin and also the subcutaneous tissues such as fat and soft tissue under the skin.

Chancre: The primary sore of syphilis that forms at the site of penetration by *Treponema pallidum*. It is a hard, nondischarging, painless lesion that erodes from the center.

Chemotherapeutic Agents: Any chemical substances used to treat diseases.

Chlamydia: Tiny, nonmotile parasitic bacteria, which are obligate intracellular parasites with a complex life cycle.

Coagulase-negative: Coagulase is an enzyme that causes coagulation particularly of blood. The coagulase test is used to differentiate the staphylococci. The coagulase-negative staphylococci, such as *S. epidermidis,* is less pathogenic than the virulent coagulase-positive *S. aureus*.

Coccobacilli: A type of bacterial cell morphology that is intermediate between a coccus and a very short bacillus.

Corolla: All the petals of a flower are called a corolla.

Corticated: Has a cortex or outer superficial layer as in part of an organism or structure.

Cyst: It is the resistant, dormant, infectious form of protozoans. It is important in the spread of pathogens such as *Giardia lamblia* and *Entamoeba histolytica*. Certain bacteria form a spherical, thick-walled cyst that is used for only survival.

D

Dark-field Microscope: A microscope in which an object such as a Spirochete is illuminated only from the sides so that the object is bright against a dark background. This procedure enhances contrast in unstained objects such as spirochetes.

Dermatophytes: These fungi cause infection of the skin, hair and nails due to their ability to utilize keratin. Dermatophytes are tinea infections.

Dermatosis: A fungal disease of the skin.

Dinoflagellates: Small microbes that are plant like Protists that usually have two flagella and may or may not have a cell wall. Dinoflagellates can cause a "Red tide".

Disinfectant: A chemical substance used on inaminate objects to destroy microbes.

E

Echovirus: An ECHO (Enteric Cytopathic Human Orphan) is a type of RNA virus that belongs to the Enterovirus genus in the Picornaviridae family. The Echovirus is one of several families of viruses that affect the gastrointestinal tract and cause skin rashes. Infection with echovirus is themost common cause of aseptic menigitis.

Ectopic Pregnancy: It is when a fertilized egg has implanted outside the uterus, generally in a Fallopian tube. It is a life-threatening condition to the mother.

Endemic: A disease that is constantly present in a geographic region.

Endospore: A resistant, dormant, structure formed inside some bacteria. The bacterial genera Bacillus and Clostridium are typical endosporeformers. The endospore is produced for survival and not for reproduction.

Eukarya: One of the three Domains of living organisms. All members are eukaryotic.

F

Febrile: A person has or shows the symptoms of fever.

Fermentation: The extraction of energy by anaerobic metabolism of substrates such as glucose or other carbohydrates into simpler, reduced metabolites. In an industrial process fermentation means any one of microbial metabolism to manufacture organic chemicals or other products such as wine.

Foraminifera: Certain species of amoebas form hard, shell-like casings called tests. The protozoa, termed foraminifera produce elaborate chalky shells of calcium carbonate. These shells can serve as depth markers for geologists drilling for oil.

Furuncle: A large, deep, pus-filled infection also called a boil.

G

Granuloma: A solid mass or nodule of inflammatory tissue containing epithelial cells, macrophages and lymphocytes. Usually a chronic inflammation in diseases such as syphilis or TB.

Gravid: Carrying fertilized eggs or developing young.

H

Halophilic: An organism that requires a salty environment in order to grow.

I

Immunoglobulins: Protein molecules that are produced by plasma cells in response to an immunogen and function as antibodies. The immunoglobulins are divided into five classes: IgG, IgM, IgA, IgD and IgE.

Impetigo: A contagious pyoderma caused by staphylococci, streptococci, or both.

Incubation Period: In the stages of an infectious disease it is the time elapsed between exposure to a pathogenic microbe and the appearance of signs and symptoms.

K

Keratin: Keratin is a type of protein found in skin, hair and nails.

L

Lactobacilli: Type of nonsporeforming gram-positive rods found in many foods used to produce cheeses, sourdough, yogurt, and other fermented foods.

M

Malaise: A general feeling of discomfort and uneasiness, the exact cause is difficult to identify. Many medical conditions can cause malaise.

N

Nematode: A name for helminths called roundworms with a long, cylindrical, unsegmented body with a heavy cuticle.

Nonhemolytic: The microbe does not produce a hemolysin, which lyses red blood cells.

O

Oxidation: The loss of electrons and hydrogen atoms by one reactant.

P

Peritonitis: An inflammation of the thin lining of the abdominal wall, which is known as the peritoneum.

Peritrichous flagellated: An arrangement of flagella, which are distributed randomly over the surface of the bacterial cell.

Phototroph: An organism that produces its own food from inorganic substances using light for energy. Green plants, certain algae, a photosynthetic bacteria are phototrophs which are also called photoautotrophs.

Picornavirus: A small RNA virus, different genera can cause polio, common cold and hepatitis.

Polarly flagellated: Flagella confer motility in procaryotes and are attached at one or both ends of the cell.

Poxviridae: Family in which the smallpox virus resides.

Protist: A diverse group of eukaryotic microbes who historically were treated as a biological kingdom called the Protista. They are unicellular and exist as independent cells. Occurring in colonies they do not show tissue differentiation. Examples of protist are algae, dinoflagellates, Euglena and Trypanosomes.

Protozoans: Single celled diverse group of microscopic, animal-like eukaryotic microbes, many of which are motile.

R

Radiolaria: They are shelled amoeba found exclusively in the sea. They live within finely sculptured glassy skeletons made of silicon dioxide. These skeletons contribute to sediment and serve as markers for oil.

Red Tide: Caused by the red pigmant produced by Dinoflagellates in marine water.

Ringworm: Ringworm, also known as dermatophytosis or tinea, is a chronic fungal infection of the skin, hair, or nails. This common name is actually a misnomer.

S

Septicemic: Systemic microbial infection associated with microbes multiplying in the blood such as in the plague.

Spirochete: Bacteria with a helical form containing a flagella (arial filaments) enclosed within a periplasmic space.

T

Taeniasis: A parasitic infection caused by the tapeworm species *Taenia soluim* (pork tapeworm) when acquired by eating raw or undercooked meat.

Taxonomic (Taxonomy): A branch of science that deals with the science of classification.

Tetrads: An arrangement of cocci in groups of four.

Tinea: Types of fungal skin infections that includes ringworm, athlete's foot and jock itch.

Toxoplasmosis: A disease caused by a sporozoan protozoan *Toxoplasma gondii.*

Trachoma: An eye disease caused by the obligate intracellular bacteria *Chlamydia trachomatis* that can cause blindness.

Trichinosis: A disease caused by trichinae in the intestines and muscle tissue of humans. It is acquired by eating raw or undercooked pork or wild game that is infected with the round worm *Trichinella spiralis.*

Trophozoite: The vegetative form of protozoans (feeding form) as opposed to a cyst or resting form.

U

Umbelliferous: Of or related to the carrot family of mostly aromatic plants with hollow stems.

V

Volar: Pertaining to the flexor surface of the wrist.

W

Whey: The liquid portion of milk resulting from bacterial enzyme production, which causes milk coagulation and separation from the solidified curd.

Wold: An upland area of open country or a rolling or hilly region.

Z

Zoonosis: A disease indigenous to animals that can be transmitted from animals to humans.

Zooplankton: The animal like component of aquatic food chains. Many species of protozoa are part of the zooplankton.

About the Author

Larry Phillip Elliott, Ph.D was born in Fleming, Missouri, on September 27, 1938 to Melvin J. Elliott and Margaret M. Elliott. He went to a one-room schoolhouse in Fleming the first semester of his first grade. The family then moved to Liberty, Missouri, where he continued his education in this system and graduated from Liberty High School in 1956. He then continued his education across town on another hill at William Jewell College, where he received his B.A. in 1960 in Biology. He received a teaching assistantship in Bacteriology at the University of Wisconsin, Madison and received a M.S. in 1962. He then received a research assistantship and finished his PhD degree in Bacteriology in 1965.

He married Wilma Lee Grove on August 25, 1961. They have three daughters: Kerrie Lynn McDaniel, PhD; Kimberly Ann Dallas M.S., R.N.; and Kelly Jo Kreis, M.D. They have six grandchildren: Christopher McDaniel, Sarah McDaniel, Carder Dallas, Aby Dallas, Audrey Kreis and Elijah Kreis.

He accepted an Assistant Professorship of Biology at Western Kentucky State College in September of 1965. In the spring of 1966 the school became Western Kentucky University (WKU). He advanced to Associate Professor and to Professorship teaching primarily Microbiology and related courses. He retired on June 30, 2002 and continued five more years in optional retirement. Thus he taught at WKU for 42 years and is currently a Professor Emeritus in the department.

During this time, he served on many University, College, and Department committees, too numerous to mention. He helped start the Mu Gamma Chapter of Tri-Beta in 1972, a national biology and honor society at WKU, and served as co-advisor for 30 years. He helped start Sigma XI at WKU, which is a national honor society for scientists and engineers, served on various committees and was elected President 1989-90. He helped start the Association of Medical Technology Students Club in 1984 and served as advisor for several years until retirement. He was a member of the University Screening Committee on Cooperative Education, which started cooperative education at WKU. He was elected Outstanding Faculty Coordinator in the state by the Cooperative Education Association of Kentucky in 1979-80. He was a member of the Steering Committee of Kentucky Outlook 2000: A strategy for Kentucky's Third Century. He was appointed a member of the Biodiversity Council of KY in 1995. He was appointed a member of the Environmental Kentucky Educational Council Master Plan Committee in Postsecondary Education in 1998. He was elected to the Governing Board for the Kentucky Academy of Science for eight years and was Vice-President, President-elect, President (1994), and Past-President. He obtained an Area Health Education System (AHES) grant in 1977, which continually funded 11-20 students which were placed in acute-care community hospitals for four-weeks until retirement in 2002. He was a Research Participant, funded by the National Science Foundation in the Department of Microbiology at the University of Iowa, Iowa City during the summer of 1970. He was on sabbatical leave in the Department of Microbiology, University of Kansas during the fall semester of 1973 doing research and taking courses in Virology and Mechanisms of Microbial Pathogenicity and Immunology.

During the summers of 1976, 1977, 1980, and 1982, he was an invited Professor at Tech Aqua Biological Station in TN where he taught Aquatic Microbiology. He received an Ogden Foundation Fellowship for Instructional Improvement and attended a Bacteriology class and did research in the Department of Microbiology at the University of Louisville during the summer of 1984.

He was Chairperson of 18 MS, 3 MA students, and 4 MSCT students who graduated. He was a member of 30 additional MS graduate student committees. He and his students presented 36 papers at scientific meetings. He or he and his students published 46 articles in scientific journals. He also published two technical (Mimeo) reports and one Protistology Laboratory Manual. He also directed the thesis of three undergraduate Honor Students in Biology.

He has been a member of First Baptist Church in Bowling Green, KY since 1965.

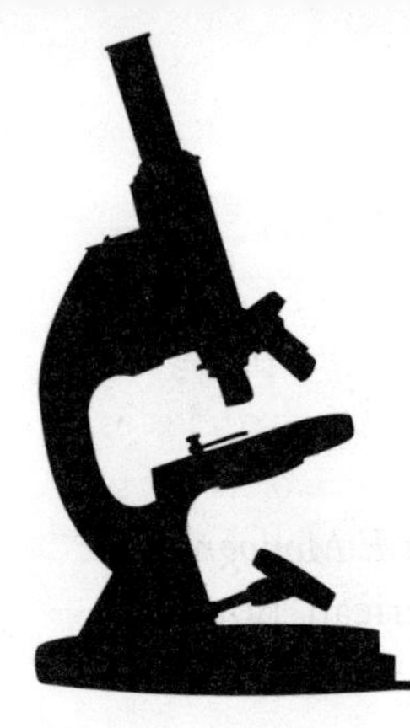

INDEX

- A -

acetic acid 26, 27
Acetobacter 26, 27
Acetobacter aceti 27
aciduric lactobacilli 30
Acinetobacter 98
Adams, John 35
aerobes 76
Ague 48
ahermatypic 40
alcohol 27
alopecia 61
amino acids 107
amoebic 53
amylase 21
anaerobes 76
anaerobic bacteria 37, 104
anethole 81
Anethum graveolens 36
Anopheles 56
anthrax 15, 46
anthrax bacillus 48
anthrax endospores 46
anthrax toxins 47
antibiotics 28
antimicrobial 7, 21, 77, 84, 86
antiseptics 81
Aphanothec halophytican 94
Aquilaria agallechum 85
Archaea 12
archaeon 12
Ascaris eggs 65
Ascaris lumbrocoides 65
ascospores 17
Aspergillus parasiticus 83
Azzouz, M.A. 87

- B -

bacillus 46, 63
Bacillus anthracis 46, 47
Bacillus subtilis 79
bacteremia 50, 55
bacteria 12, 30, 32, 35, 38, 50, 52, 55, 57, 59
Bacteria, Lactic Acid 34
bacteriocins 101
bacteriology 6
bacteriostatic 83
Bacteroides 99
basidiomycete 42
basidiospores 43
Batzing, B.L. 15
benzaldehyde 78
beta-hemolytic streptococci 54
Bifidobacterium 99
biochemical 26, 33
Biology 25
Blumenthal, M. 87
B. melitensis 50
Botryococcus braunii 105
Brassica hirta 82
Brassica nigra 82
bromide 93
bronchial asthma 80
bronchitis 80
Brucella 50
brucellosis 15, 49, 73
Bruce, Sir David 50
B. suis 50
buboes 51
bubonic plague 15, 51
Bullerman, L.B. 87
Bullerman, L.R. 87
Burhans, Dr. Rollin 6

- C -

Caesar, Julius 35
calcium 40
calcium carbonate 40
calcium chlorides 93
carbohydrate 26
carbon dioxide 17, 21, 23, 76
cardiovascular 72
Carr, F.J. 34
cell 17, 27, 54
cell proteins 78
cell wall 28

chancre 58
chemolithotrophic bacteria 107
Chill, D. 34
Chlamydia trachomatis 48
Chlymdia trachomatis 57
chronic amebiasis 53
Cieslak, T.J. 73
cinnamic aldehyde 83
Clauson, Mark 6
Clostridium 99
Clostridium pasteurianum 38
Clostridium species 38
coagualse 48
coagulant 78
coagulates 48
cocci in chains 30
coccobacilli 50
coccus 32
Codex Ebers 79
Columbus, Christopher 35, 38
consortia 38
Corallium 39
Corallium japonicum 41
Corallium rubrum 39, 41
Coriandium sativum 85
corynebacteria 48, 98, 100
Cowan, M.M. 86
cremoris 34
Cucumis chate 35
cutaneous anthrax 47
cutaneous mycosis 61
Cyclops 66
cysticerosis 73
Cysts 54
cytologic 59

- **D** -

Dallas, Aby Jewell 6
Dallas, Carder 6
Dallas, Kimberly Ann Elliott 6
Dallas, Rett 6
denaturizer 78
dermatomycosis 61
dermatophytes 61, 76
dermatoses 60
dextrose 77
DeYoung 11
digesting enzymes 21
dinoflagellates 70
diplococcus 57
Dracunculiasis 66
Dracunculus 66
Dracunculus medinensis 66
Dunaliella salina 94
Dye, D.L. 13
dysentery 80

- **E** -

eczema 68, 80
Eerdmans, William B. 13
effervescence 26
Elliott, Larry P. 7
Elliott, Margaret Marie 6
Elliott, Melvin J. 6
Elliott, Wilma 6
E. M. Eitzen 73
endocarditis 50
endospore 46
Entamoeba histolytica 52
Enterococcus 35
Enterococcus faecalis 99, 100
Entrobacter 35
enzyme 21, 37, 38, 48, 79
Epidermophyton 61
epithelial 57
Erysipelas 54
eschar 47
Escherichia coli 79, 83, 84, 99
estragol 81
ethanol 23, 26, 78
ethyl alcohol 17, 26
eucaryotic 53, 67
eucaryotic cells 97
eucaryotic microbes 97
eucaryotic molds 61
eugenol 83
Eukaryotic cells 12
eukaryotic microbes 38
extremophilic 94

- **F** -

Facultative anaerobic 53
fermentation 17, 21, 26, 35
fermented juice 18
Francisella tularensis 72
Frazier, W.C. 31, 86
fructose 16
fulminating 46
fungus 12, 17, 42, 61
furanno-coumarins 81
Fusobacterium 99

- **G** -

gastrodermal cells 40
gastrointestinal disease 67
germinated 21
globules coalesce 34
Gluconobacter 26, 27
glucose 16
glucoside sinigrin 82
glycoside 30
Gonococcal pharyngitis 57
gonococcus 57
Gonorrhea 57
Gram, H.C. 28
gram-negative 28, 51, 57, 76
gram-negative bacteria 35
gram-positive 30, 32, 76

gram-positive bacterium 46
Gram stain 27
granules 34
granulomas 50
guinea worms 66
gummas 59

- H -

Haemophilus 98
Haemophilus influenzae 69
Halobaculum 94
Haloferax 94
halophiles 13, 76
halophilic Archae 94
halophilic Archaea 93
Halorubrum 94
Hammer, J. 94
Hansen's bacillus 64
Harley, J.P. 19
Helicobacter pylori 78
hermatypic 40
hieroglyph 66
high osmolarity 77
histologic reactions 59
histoplasmosis 15
Human Immunodeficiency Virus 56
Humulus lupulus 21
hydrocarbons 105
hydrogen peroxide 77
hydrolyze 30
hydrothermal vents 12
hypersaline 94
hypersaline basin 93
hyphae 43, 61
hypopigmented 63
Hyssopus officinalis 84

- I -

immunodeficiency virus 91
impetigo 49
inflammatory cell 48
influenza 47
iodine 27

- J -

Jenner, Edward 69
Jesus Christ 6, 7

- K -

Klein, D.A. 19
Koch, Robert 45
Kries, Audry 6
Kries, Eli 6
Kries, Kelly Jo Elliott 6
Kries, Shawn 6

- L -

L. acidophilus 32
lactic acid 32, 34
lactic acid bacteria 30, 36, 102
lactobacilli 32, 98, 100
Lactobacillus 34
Lactobacillus brevis 35
Lactobacillus casei 32
Lactobacillus lactis 34
Lactobacillus mesenteroides 29
Lactobacillus plantarum 30, 35
Lactococcus lactis 32
L. bulgaricus 32
Lederberg, J. 13
Leeuwenhoek, Anton van 12, 45
Lepromatous leprosy 63
leprosy bacillus 63
Leuconostoc mesenteroides 30
Leuconostsoc mesenteroides 35
levulose 77
Lieu, F.Y. 87
Linum usistatissimum 37
Listeria monocytogenes 84
locomotor ataxia 59
Lumbricoides 65
Lusken, Tovis 104
lye 30
lyhmphatic 58

- M -

Madigan, M.T. 25
Madison, Dolly 35
magnesium 78, 93
Maida, N. 34
malaria 48
Mandragora 81
Martinko, J.M. 25
McDaniel, Christopher 6
McDaniel, Kerrie Lynn 6
McDaniel Sarah 6
Meningitis 69
Mentha piperita 81
Menthol 81
methane 12
methanogens 12
methyl chavicol 81
microbe 6, 7, 12, 14, 16, 36, 42, 45, 47, 48, 61, 72, 74, 75, 76, 77, 78, 91, 108
microbial cells 76
microbial diseases 15, 46
microbial entry 45
microbial infections 56
microbial invasion 45
microbiology 6, 7, 19, 27, 31, 86
microcrustacean 66
microflora 30
microorganisms 25, 32, 33, 47, 74, 107
Microsporum 61
monotonous 68
mucosa 57
mucosal membranes 58
mucous membranes 68
multilateral budding 17
musculoskeletal system 58
mycelia 17, 61

mycelium 43
mycobacteria 63
Mycobacterium 71
Mycobacterium leprae 62
Mycobacterium tuberculosis 71, 108
mycorrhizae 108
mycorrhizal 108
mycotoxigenic Aspergillus 83
myrosin 82

- N -

nasopharynx 98
Neisseria 98
Neisseria meningitidis 69
neutrophils 57
Nigella sativa 80
niter 26
nonmotile 94
nucleic acid 67
nucleocapsid 67
nucleus 12

- O -

oleuropein 30
organism 12, 32, 33, 57
Origanum maru 84
Origanum vulgare 84
oxidation 26
oxidize 26
oxygen 12, 35

- P -

Palsy 69
paralysis 69
Parker, J. 25
Pasteur, Louis 79
pathogens 42
pectinase 37
Pediococcus cerevisiae 30, 35
peptic ulcers 78
Peptostreptococcus 99
peritrichous flagella 59
pH 30, 36, 77, 93, 101
Phagocytes 57
phagocytized 57
pharynx 57
picornaviruses 67
Pimpinella anisum 80
plague septicemic 51
plant molecular biology 6
Plasmodium 55
plasmolyhsis 76
Plasmopara viticola 42
pneumonia 72
pneumonic 51
pneumonic tularemia 72
pneumonitis 65
poliomyelitis 66
Polioviruses 67
potash 93
Prescott, L.M. 19
procaryotic 67, 97
prokaryotes 12
Propionibacterium 98
proteolytic enzymes 54
protozoa 12, 55, 105
protozoans 53
Pseudomonas aeruginosa 84
Pulmonary anthrax 47
Pyoderma 48

- R -

rabies 15
Radiolarian fossil 101
rheumatism 80
Rhizopus 43, 44
Rhizopus stolonifer 44
ribosomal ribonucleic acid 93
Rosenberg, Joel C. 104, 106
Ruta graveolens 84

- S -

Saccharomyces cerevisiae 23, 79
Sacchromyces carlsbergensis 21
Sacchromyces cerevisiae 21, 26
Sacharomyces cerevisiae var. ellipsoideus 16
safranin 27
salarium argentium 74
Salmonella typhi 59
Sarcoptes scabiei 60
sarcoptic mange 60
S. dysenteriae 53
septicemia 46, 72
Seren, S.A. 87
Shfela Basin 105
Shigella 52
Shigella dysenteriae 53
shigellosis 53
Smallpox 68
Sodium chloride 74
spinal meningitis 69
spirochetes 58
spoilage microorganisms 33
S. pyogenes 55
S. sonnei 53
staphylococcus 48, 98, 100
Staphylococcus aureus 48, 98
Staphylococcus epidermidis 100
Stapyhylococcus aureus 84
Streptococcal impetigo 49
streptococci 48, 98, 100
Streptococcus pneumoniae 69
Streptococcus pyogenes 49, 54
Streptococcus sanguis 98
streptomycin 52
sulfoxide 79
syphilis 68
syphilitic 58

- T -

taeniasis 73
taxonomic 28
tenesmus 52
tetracycline 52
tetrads 31
Thomas, Olive 6
Thomson, Cole 15
tinea 68
tinea barbae 61
tinea capitis 61
tinea corporis 61
tinea cruris 61
tinea unguium 61
toxoplasmosis 73
Treponema pallidum 57, 62
trichinosis 73
Trichophyton 61
Trichophyton schoenleinii 61
Trophozoites 54
tuberculoid leprosy 63
tuberculosis 52, 62, 70
Tyler, V.E. 86
Typhoid fever 60

- U -

ulceroglandular tularemia 72
Umbelliferae 80
unicellular 17, 40
unicellular microbes 56
Unicinula necator 42
usitatissimum 38

- V -

Variola Minor 68
vinegar bacteria 26
viridans streptococci 98

- W -

Washington, George 35
Westhoff, D.C. 31, 86

- Y -

Yeast cells 17
yeasts 7
Yersenia enterocolitica 84
Yersinia pestis 51

- Z -

zoonosis 46, 49
zoonotic diseases 14
zooxanthellae 40

Scriptural References

Scripture references appear bold, page numbers appear italic.

Old Testament

Genesis

2:4b • *pp 11*; **3:7** • *pp 77*; **6:1-9:2** • *pp 14*; **7:11** • *pp 107*; **7:16** • *pp 14*; **8:2** • *pp 107*; **8:11** • *pp 29*; **9:20** • *pp 16*; **9:21-24** • *pp 18*; **14:3** • *pp 93*; **14:10** • *pp 103*; **17:10-14** • *pp 91*; **18:6** • *pp 23*; **18:8** • *pp 34*; **19:3** • *pp 24*; **19:31-38** • *pp 18*; **21:4** • *pp 91*; **27:9, 14, 17** • *pp 107*; **29:17** • *pp 47*; **30:14-17** • *pp 82*; **32:14** • *pp 107*; **35:8** • *pp 108*; **43:11** • *pp 76*; **49:25** • *pp 103*; **50:1-3** • *pp 88*; **50:2** • *pp 89*; **50:2, 3, 26** • *pp 88*; **50:12** • *pp 89*; **50:23-26** • *pp 88*; **Chapter 1** • *pp 11*

Exodus

3:8 • *pp 76*; **4:11** • *pp 45*; **7:20-21** • *pp 70*; **8:13** • *pp 46*; **9:1-12** • *pp 46, 48*; **9:8-12** • *pp 68*; **9:31** • *pp 37*; **12:8** • *pp 24*; **12:15** • *pp 24*; **12:15-20** • *pp 24*; **12:18-20** • *pp 24*; **12:22** • *pp 83*; **12:34** • *pp 23, 24*; **12:48** • *pp 91*; **15:25-26** • *pp 89*; **15:26** • *pp 89, 97*; **16:3** • *pp 86*; **16:4** • *pp 75, 95*; **16:23-30** • *pp 95*; **16:31** • *pp 36, 76*; **16:32-36** • *pp 95*; **21:18-19** • *pp 89*; **23:11** • *pp 29*; **27:20-21** • *pp 24*; **30:23-25** • *pp 83*; **34:7** • *pp 57*; **34:22** • *pp 24*; **39:10-13** • *pp 40*

Leviticus

5:2 • *pp 91*; **6:17** • *pp 24*; **7:12** • *pp 24*; **7:13** • *pp 24*; **10:9** • *pp 18*; **10:19** • *pp 20*; **11:6-8** • *pp 72*; **11:7-8** • *pp 73*; **11:24-40** • *pp 91*; **12:1-8** • *pp 91*; **12:3** • *pp 91*; **13:1-17** • *pp 62*; **13:2** • *pp 49*; **13:18** • *pp 49*; **13:28** • *pp 49, 59*; **13:29** • *pp 61*; **13:30** • *pp 68*; **13:42** • *pp 61*; **13:47-50** • *pp 43*; **14:1-32** • *pp 91*; **14:4** • *pp 83*; **14:4-7** • *pp 62*; **14:33, 57** • *pp 91*; **14:49-52** • *pp 84*; **14:54-57** • *pp 43*; **15:4** • *pp 56*; **15:9** • *pp 56*; **20:24** • *pp 76*; **21:1-4** • *pp 91*; **21:20** • *pp 60, 61, 68*; **22:22** • *pp 61*; **23:17** • *pp 24*; **24:1-2** • *pp 30*; **24:1-23** • *pp 24*; **26:14-16** • *pp 89*; **26:14-25** • *pp 108*; **26:16** • *pp 48, 52, 55, 72*; **26:26** • *pp 25*; **26:29** • *pp 68*; **Chapter 13** • *pp 49, 91*; **Chapter 14** • *pp 49*; **Chapter 15** • *pp 49, 56*

Numbers

5:2 • *pp 91*; **5:2-3** • *pp 91*; **5:15** • *pp 22*; **6:3** • *pp 18, 20, 26*; **9:6, 10** • *pp 91*; **10:29** • *pp 18*; **11:5** • *pp 35, 74, 78*; **11:6** • *pp 95*; **11:7** • *pp 36, 86*; **13:20-24** • *pp 16*; **15:20-21** • *pp 25*; **16:13** • *pp 76*; **19:2-6** • *pp 84*; **19:6** • *pp 83*; **19:11-22** • *pp 91*; **21:6** • *pp 65*; **22:24** • *pp 16*; **24:6** • *pp 85*; **31:19** • *pp 91*; **34:3** • *pp 93*; **34:12** • *pp 75*

Deuteronomy

3:17 • *pp 75, 93*; **4:19** • *pp 93*; **6:11** • *pp 29*; **7:12-16** • *pp 89*; **8:3** • *pp 25*; **8:8** • *pp 22, 29, 76*; **14:8** • *pp 73*; **14:23:26** • *pp 17*; **15:14** • *pp 17*; **18:4** • *pp 17*; **22:11** • *pp 37*; **23:1** • *pp 90*; **23:13-14** • *pp 91*; **24:8** • *pp 91*; **24:20** • *pp 30*; **28:22** • *pp 43, 54, 59, 72*; **28:27** • *pp 52, 60, 68*; **28:35** • *pp 90*; **28:40** • *pp 29*; **28:60** • *pp 51*; **29:6** • *pp 20*; **29:23** • *pp 75*; **32:13** • *pp 76*; **32:14** • *pp 33*; **32:39** • *pp 45, 89*; **33:13** • *pp 103*; **33:19** • *pp 103, 105*; **33:24** • *pp 30, 103*

Joshua
2:1 • *pp 37*; **2:6** • *pp 37*; **5:12** • *pp 95*; **9:4, 13** • *pp 17*; **9:5, 12** • *pp 43*

Judges
1:16 • *pp 18*; **4:19** • *pp 34*; **5:25** • *pp 34*; **9:8** • *pp 29*; **9:10-11** • *pp 77*; **12:7, 24** • *pp 18*; **13:4** • *pp 20*; **13:4, 7, 14** • *pp 18*; **13:7** • *pp 20*; **13:24** • *pp 20*; **14:8** • *pp 76*

Ruth
2:14 • *pp 26*

1 Samuel
1:24 • *pp 17*; **5:6, 9, 11** • *pp 51*; **5:6, 9, 12** • *pp 52*; **6:7** • *pp 107*; **9:7** • *pp 25*; **14:25** • *pp 76*; **16:20** • *pp 25*; **17:17-18** • *pp 33*; **17:18** • *pp 33*; **21:3-6** • *pp 24*; **25:18** • *pp 17, 77*; **25:36-37** • *pp 18*

2 Samuel
8:13 • *pp 75*; **12:1-4** • *pp 107*; **16:1** • *pp 17, 25*; **16:2** • *pp 17*; **17:28** • *pp 107*; **17:29** • *pp 33, 34*

1 Kings
4:33 • *pp 84*; **6:31-33** • *pp 29*; **8:37** • *pp 43*; **13:4-6** • *pp 89*; **14:3** • *pp 76*; **17:17-22** • *pp 89*; **20:12-16** • *pp 18*

2 Kings
2:20, 21 • *pp 75*; **4:18-20, 34-35** • *pp 89*; **4:42** • *pp 23*; **5:1** • *pp 62*; **5:9-14** • *pp 62*; **5:10** • *pp 90*; **7:3-5** • *pp 62*; **10:15-23** • *pp 18*; **14:7** • *pp 75*; **18:32** • *pp 76*; **19:7** • *pp 59*; **20:7** • *pp 49, 78, 89*

1 Chronicles
2:55 • *pp 18*; **9:32** • *pp 24*; **18:12** • *pp 75*; **23:29** • *pp 24*

2 Chronicles
2:10 • *pp 30*; **6:28** • *pp 43*; **9:9** • *pp 86*; **16:12** • *pp 90*; **16:14** • *pp 88*; **20:1-30** • *pp 93*; **21:19** • *pp 53*

Nehemiah
8:15 • *pp 30*; **13:15** • *pp 16*

Esther
1:7 • *pp 17*; **1:7-10** • *pp 18*

Job
1 • *pp 89*; **2:7** • *pp 45, 49*; **6:6** • *pp 74*; **10:10** • *pp 33*; **13:4** • *pp 90*; **15:33** • *pp 30*; **20:17** • *pp 34*; **22:7** • *pp 25*; **28:18** • *pp 40*; **38:16** • *pp 107*

Psalms
14:4 • *pp 25*; **33:12** • *pp 108*; **38:1-8** • *pp 56*; **38:5** • *pp 68, 90*; **39:11** • *pp 68*; **45:8** • *pp 85*; **52:10** • *pp 31*; **68:30** • *pp 107*; **69:21** • *pp 26*; **78:66** • *pp 52*; **80:10** • *pp 16*; **81:16** • *pp 76*; **104:15** • *pp 17*

Proverbs
3:15 • *pp 40*; **7:17** • *pp 83, 85*; **8:11** • *pp 40*; **9:5** • *pp 17*; **10:26** • *pp 26*; **11:22** • *pp 73*; **12:4** • *pp 90*; **14:30** • *pp 90*; **16:24** • *pp 76*; **20:1** • *pp 18, 20*; **20:15** • *pp 40*; **21:17** • *pp 18*; **23:20** • *pp 18*; **23:29-35** • *pp 18*; **24:13** • *pp 76*; **25:20** • *pp 26*; **28:7** • *pp 20*; **30:15** • *pp 89*; **30:33** • *pp 33, 34*; **31:6** • *pp 20, 90*; **31:6-7** • *pp 17*; **31:10** • *pp 40*; **31:13** • *pp 37*

Ecclesiastes
9:7 • *pp 17*; **10:19** • *pp 17*

Song of Solomon
4:11 • *pp 76*; **4:14** • *pp 85*; **5:1** • *pp 76, 86*; **6:2** • *pp 86*; **7:2** • *pp 17*; **7:13** • *pp 82*; **8:2** • *pp 17*

Isaiah
1:6 • *pp 49, 90*; **1:8** • *pp 16, 35*; **3:7** • *pp 89*; **3:17** • *pp 68*; **5:11** • *pp 20*; **5:11-12** • *pp 18*; **7:14-15** • *pp 33*; **7:15,22** • *pp 76*; **17:6** • *pp 30*; **19:9** • *pp 37*; **24:13** • *pp 30*; **28:1** • *pp 18*; **28:21** • *pp 78*; **28:25, 27** • *pp 80*; **28:27** • *pp 80*; **29:9** • *pp 20*; **38:21** • *pp 49*; **43:17** • *pp 37*; **58:7** • *pp 25*; **63:2** • *pp 16*; **63:3** • *pp 16*

Jeremiah
7:18 • *pp 23*; **8:22** • *pp 89, 90*; **11:16** • *pp 29*; **13:12** • *pp 17*; **19:29** • *pp 84*; **20:15** • *pp 75*; **25:30** • *pp 16*; **35:5-6, 14** • *pp 18*; **35:6** • *pp 18*; **37:21** • *pp 25*; **41:8** • *pp 76*; **46:11** • *pp 89*; **48:33** • *pp 16*; **51:8** • *pp 90*

Ezekiel
4:9 • *pp 21, 107*; **6:13** • *pp 108*; **16:13** • *pp 76*; **18:7** • *pp 25*; **24:23** • *pp 68*; **27:6** • *pp 108*; **27:16** • *pp 40*; **27:17** • *pp 76*; **28:13** • *pp 40*; **28:17-20** • *pp 40*; **30:21** • *pp 89*; **33:10** • *pp 68*; **44:21** • *pp 18*; **45:21** • *pp 24*; **47:11** • *pp 93*; **47:18** • *pp 93*

Daniel
1:8 • *pp 18*; **1:12-16** • *pp 18*; **10:3** • *pp 18*

Hosea
2:9 • *pp 37*; **4:11** • *pp 18*; **4:13** • *pp 108*; **14:6** • *pp 29*

Amos
2:8 • *pp 17, 18*; **2:9** • *pp 108*; **4:9** • *pp 43*; **6:6** • *pp 18*; **9:13** • *pp 16*

Micah
2:11 • *pp 20*; **6:15** • *pp 30*

Haggai
2:17 • *pp 43*

Zechariah
9:17 • *pp 17*; **11:11b** • *pp 59*; **11:17** • *pp 59*; **14:12** • *pp 68*

New Testament

Matthew
3:4 • *pp 76*; **4:3** • *pp 25*; **4:4** • *pp 25*; **5:13** • *pp 74*; **7:6** • *pp 73*; **8:1-17** • *pp 62*; **8:3** • *pp 62*; **8:6** • *pp 69*; **8:14** • *pp 55*; **9:2** • *pp 69*; **9:12** • *pp 90*; **9:17** • *pp 17*; **12:10-13** • *pp 66*; **13:31** • *pp 82*; **17:20** • *pp 82*; **21:33** • *pp 16*; **23:23** • *pp 36, 80, 81*; **23:24** • *pp 17*; **25:3** • *pp 30*; **26:29** • *pp 17*; **27:34** • *pp 26*

Mark
1:6 • *pp 76*; **2:22** • *pp 17*; **3:1-5** • *pp 66*; **4:31** • *pp 82*; **5:26-27** • *pp 90*; **9:17** • *pp 45*; **9:25** • *pp 45*; **9:50** • *pp 74*; **12:1** • *pp 16*; **15:23** • *pp 26*; **15:36** • *pp 26*

Luke
1:15 • *pp 18, 20*; **1:59** • *pp 91*; **2:21** • *pp 91*; **2:22** • *pp 91*; **4:23** • *pp 90*; **4:38** • *pp 55*; **4:39** • *pp 55*; **5:13** • *pp 62*; **5:31** • *pp 90*; **5:37-38** • *pp 17*; **6:3-4** • *pp 24*; **6:6-10** • *pp 66*; **8:43** • *pp 89*; **10:34** • *pp 17, 30, 78, 90*; **11:5-6** • *pp 25*; **11:42** • *pp 81, 85*; **13:19** • *pp 82*; **17:6** • *pp 82*; **17:11-19** • *pp 62*; **24:42** • *pp 76*

John
2:7-11 • *pp 17*; **5:2-9** • *pp 66*; **6:48-58** • *pp 25*; **7:22-24** • *pp 91*; **9:1** • *pp 57*; **9:2** • *pp 45*; **19:29-30** • *pp 26*; **19:39** • *pp 85*; **19:39-40** • *pp 88*; **49-53** • *pp 55*

Acts
7:8 • *pp 91*
12:21-23 • *pp 64*
28:1-10 • *pp 49*
28:8 • *pp 52*

Romans
11:16 • *pp 25*; **11:17-24** • *pp 30*; **14:21** • *pp 19*

1 Corinthians
5:6 • *pp 25*

Galatians
5:9 • *pp 25*

Ephesians
5:3 • *pp 56*

Colossians
4:6 • *pp 74*; **4:14** • *pp 90*

1 Timothy
5:23 • *pp 17, 78, 90*; **23:8** • *pp 18*

Titus
2:3 • *pp 18*

Hebrews
9:4 • *pp 95*; **9:19** • *pp 84*

Revelation
6:7-8 • *pp 108*; **6:13** • *pp 77*; **14:19-20** • *pp 16*; **18:13** • *pp 83*; **19:15** • *pp 16*; **21:19-20** • *pp 40*

Other Books by Acclaim Press

The Gift of Creation reveals the splendor of nature in its varied landscapes, flora and fauna. The biblically-based essays remind us to cherish and care for God's great gift.

A pictorial walking tour of the Shaker Village during every season and nearly every time of day, a lasting testament to the Shaker craftsmanship.

A gathering and compilation of the teachings of many pastors, Bible scholars and historians about this wonderful book in the Bible, systematically breaking it down chapter-by-chapter, passage-by-passage, showing that Jesus Christ meant for it to be understood.

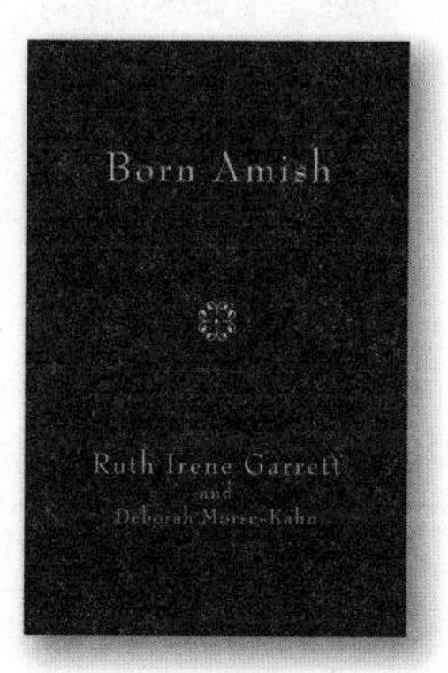

One woman's life growing up Amish to young womanhood, fallingin love with an Englischer and leaving the Amish community to begin a new life out in the world.

An inspirational book that provides strength, faith and encouragement to endure the trials of life, and to understand that since events of the past can't be changed, it is best to reflect on these events and apply the lessons learned for a better tomorrow.